The Illustrated Guide to Film Formats

Written and illustrated by
Ashley Blewer

The Illustrated Guide to Film Formats

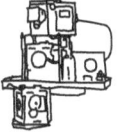
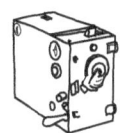

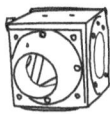

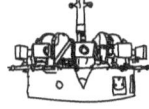
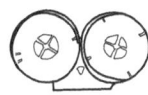

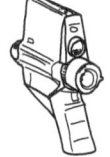

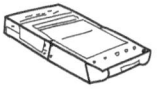

Written and illustrated by
Ashley Blewer

To my fellow honeybees, Marleigh and Travis

Table of Contents

Introduction

This book provides an introductory overview to thirty-six historically significant film formats. These formats include major developments in moving image cameras, projectors, and the medium of film.

Each format page includes the following details:

AKA: Any other notable names that this format was "also known as"

Format: Whether the format was primarily analog or digital

Developed by: The primary developer, manufacturer, or patent holder(s) for the format

Era: The approximate era of production (acknowledging that end dates are particularly fuzzy, as some formats continued to be used long past their production expiration date)

Aspect ratio: The ratio of width to height per film frame (as measured on the frame, not as projected, unless otherwise noted)

Size: The width of the film, in millimeters

Fun facts: Three informational sentences about the format

I hope you enjoy this illustrated guide and I encourage you to explore each of the formats more thoroughly on your own!

Magic lantern

Developed by
Unknown

Era
1600s

Fun fact
This was one of the first uses of moving images mostly by simple manual movements between two frames

AKA
Laterna magica, Utsushi-e

Format
analog

Fun fact
This format was a significant inspiration for what would become the moving image projector

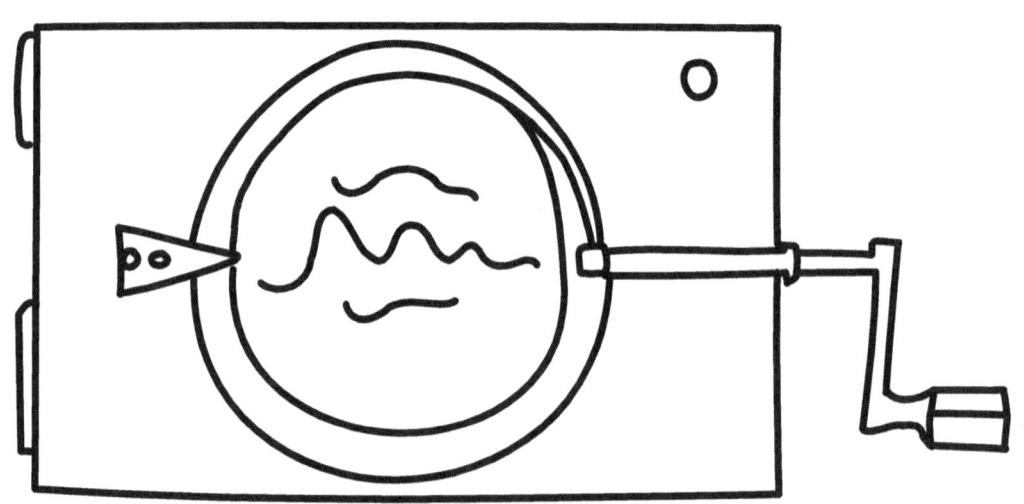

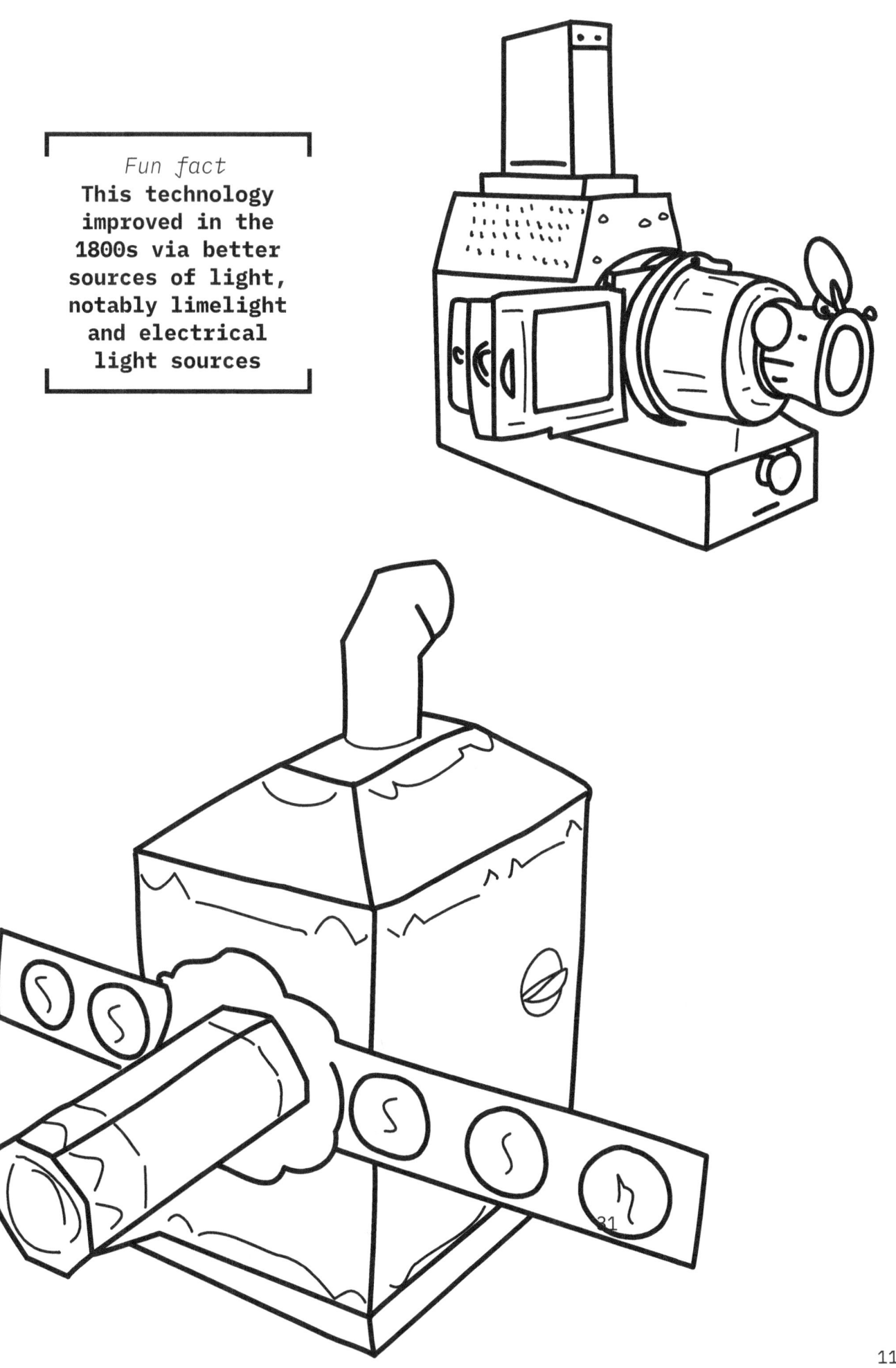

Zoopraxiscope

AKA
Zoographiscope,
Zoogyroscope

Era
1879–1900s

Fun fact
This format took
inspiration from
zoetropes (spinning
cylinders with
thin viewing windows)
and phenakistiscopes
(spinning discs) to
create the illusion
of movement

Developed by
Eadweard Muybridge

Format
analog

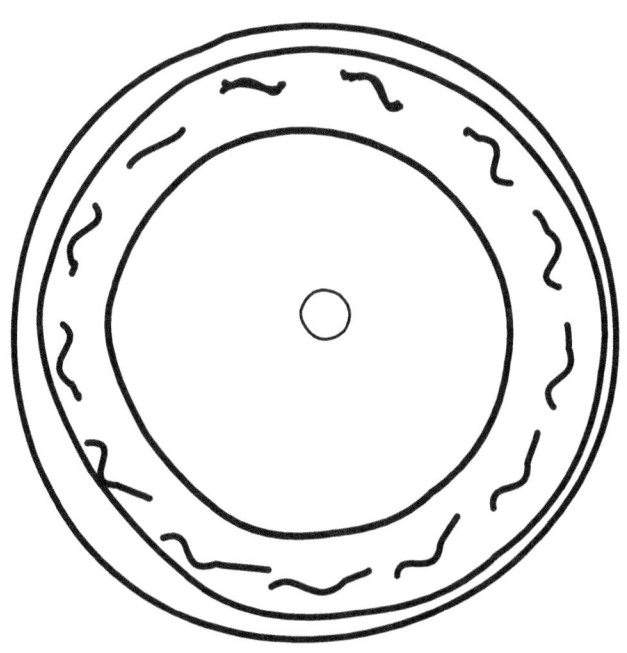

Fun fact
**This format's
illustrations were
based on motion
photography
and were painted
on glass, allowing
the projection of
realistic movement**

Fun fact
**This format inspired moving
image camera technology**

35mm (Silent)

Format
analog

Developed by
William Kennedy Dickson & Thomas Edison (Edison Company)

AKA
Silent, Edison size

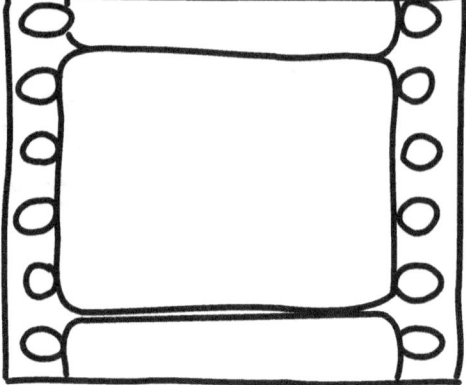

Era
1889-

Size
35mm

Aspect Ratio
1.33:1 (4:3)

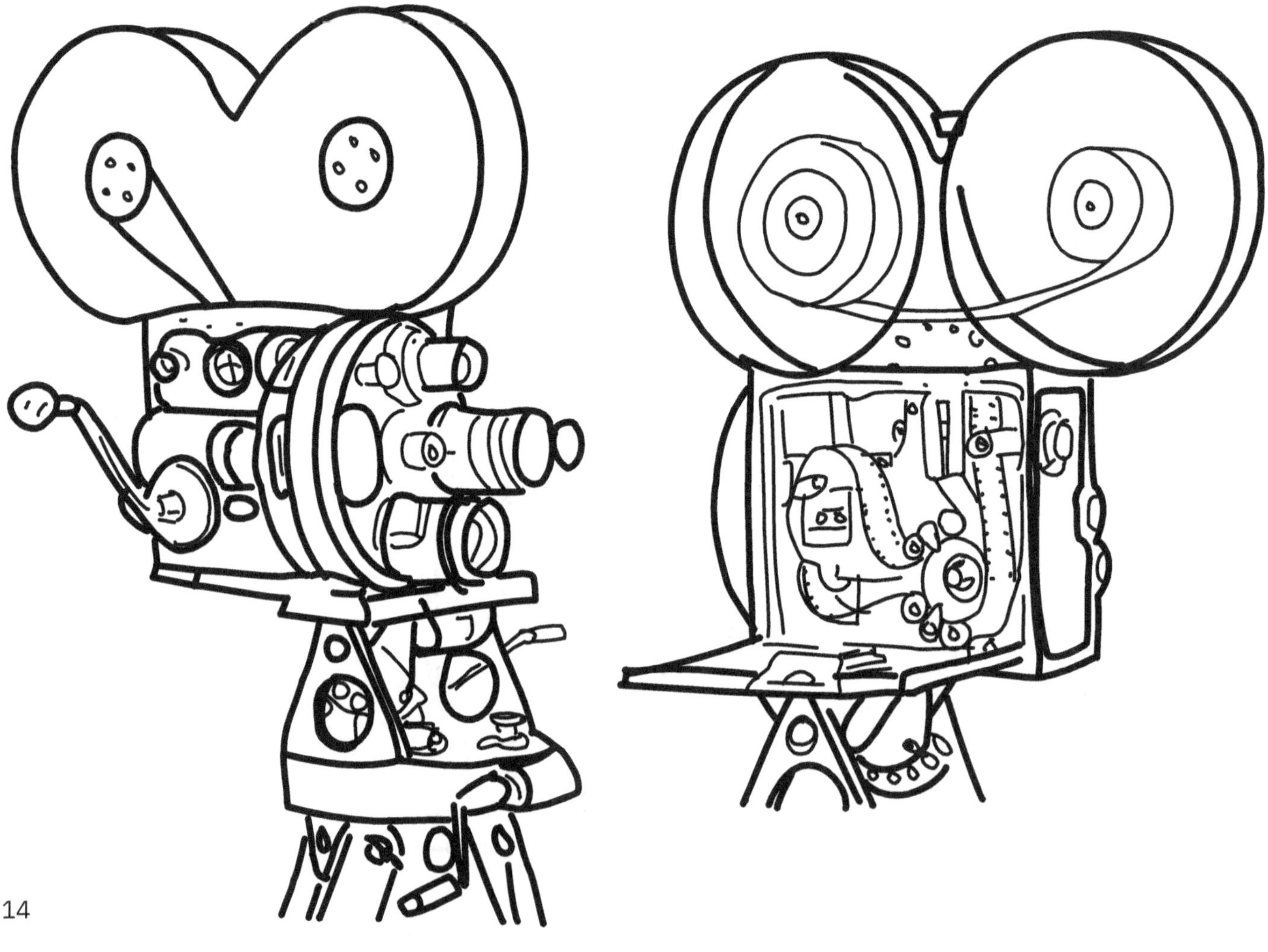

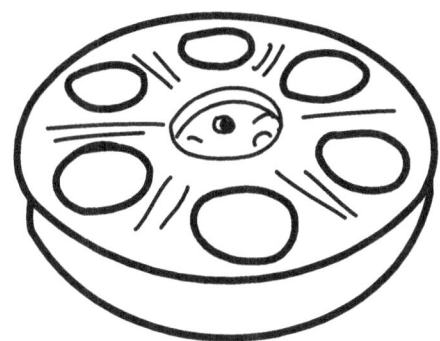

Fun fact
**This format results in
16 frames per foot of film**

Fun fact
**This format
was typically
in black & white
(unless hand-
painted)
and used
nitrocellulose
as a base**

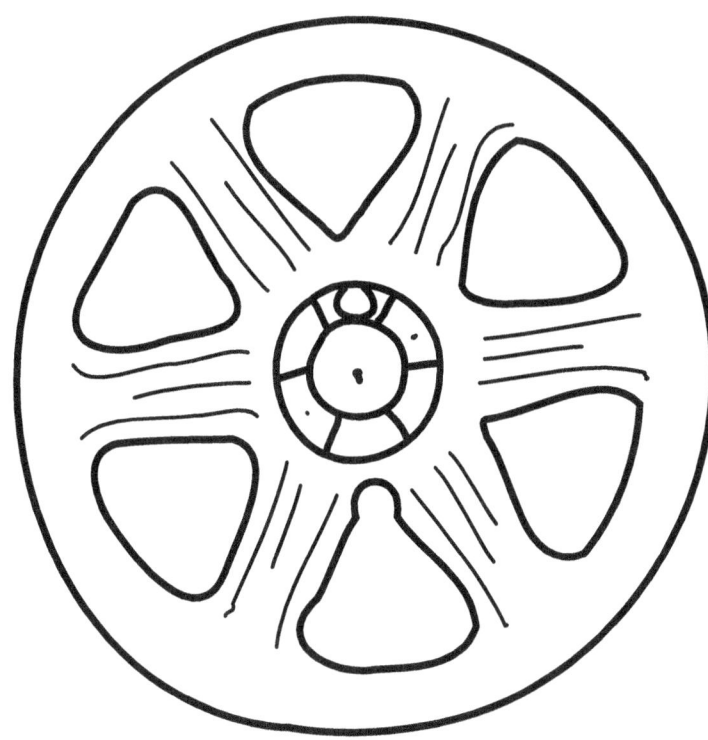

Fun fact
**This is the
most commonly
used film gauge,
introduced
around 1889,
became an
international
standard in
1909, and
remained the
dominant
moving image
film gauge
(with variations
on aspect ratio)**

Phantoscope

Format
analog

Developed by
Charles Francis Jenkins
& Thomas Armat
(Edison Company)

Era
1890

Aspect Ratio
1.33:1 (4:3)

Size
35mm

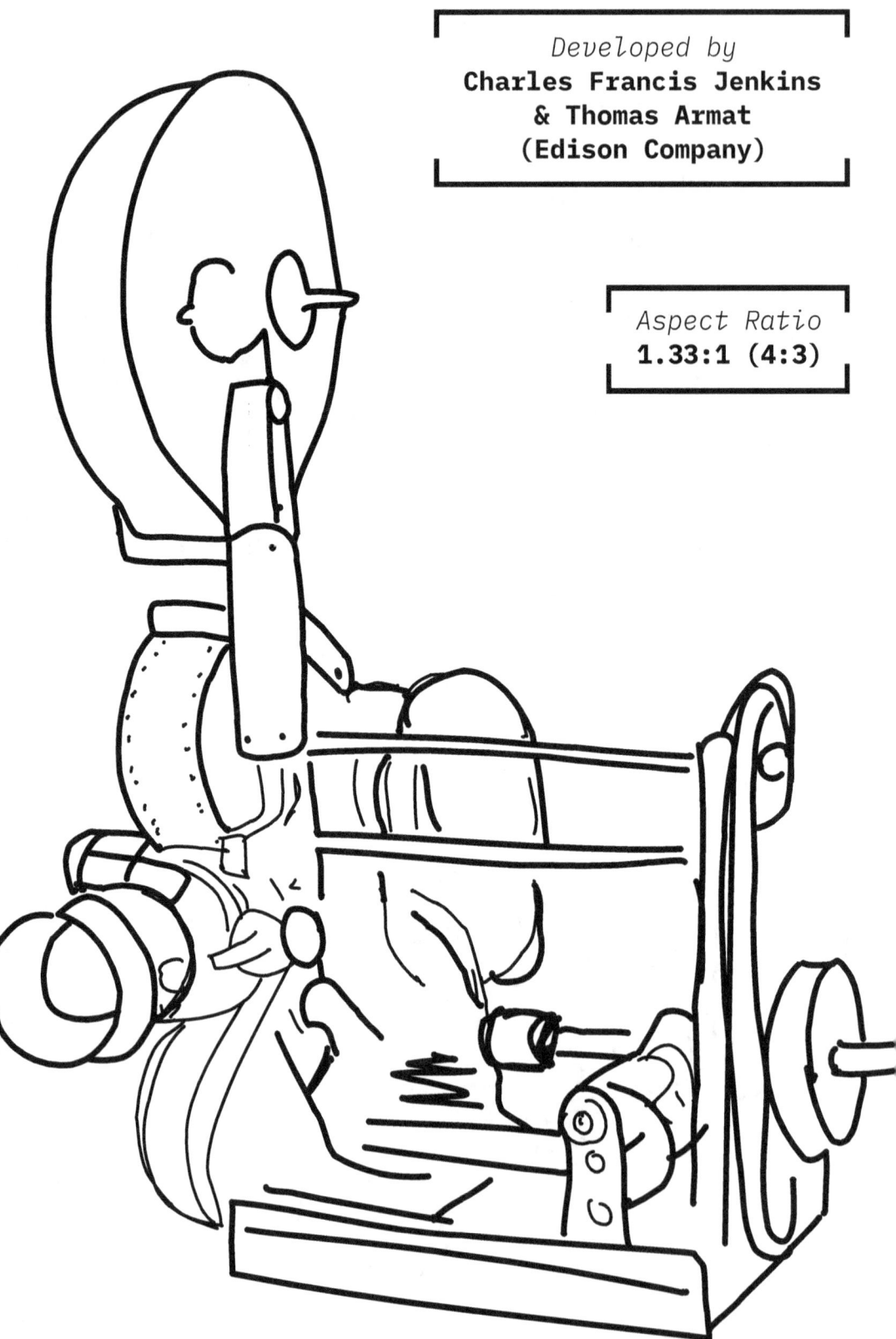

This was one of the first projectors ever created and the first projector to use 35mm film

This format should not be confused with the Phantascope, which was a motion magic lantern developed around the same time

Through a deal with the Edison Company, this format would be marketed as the Vitascope

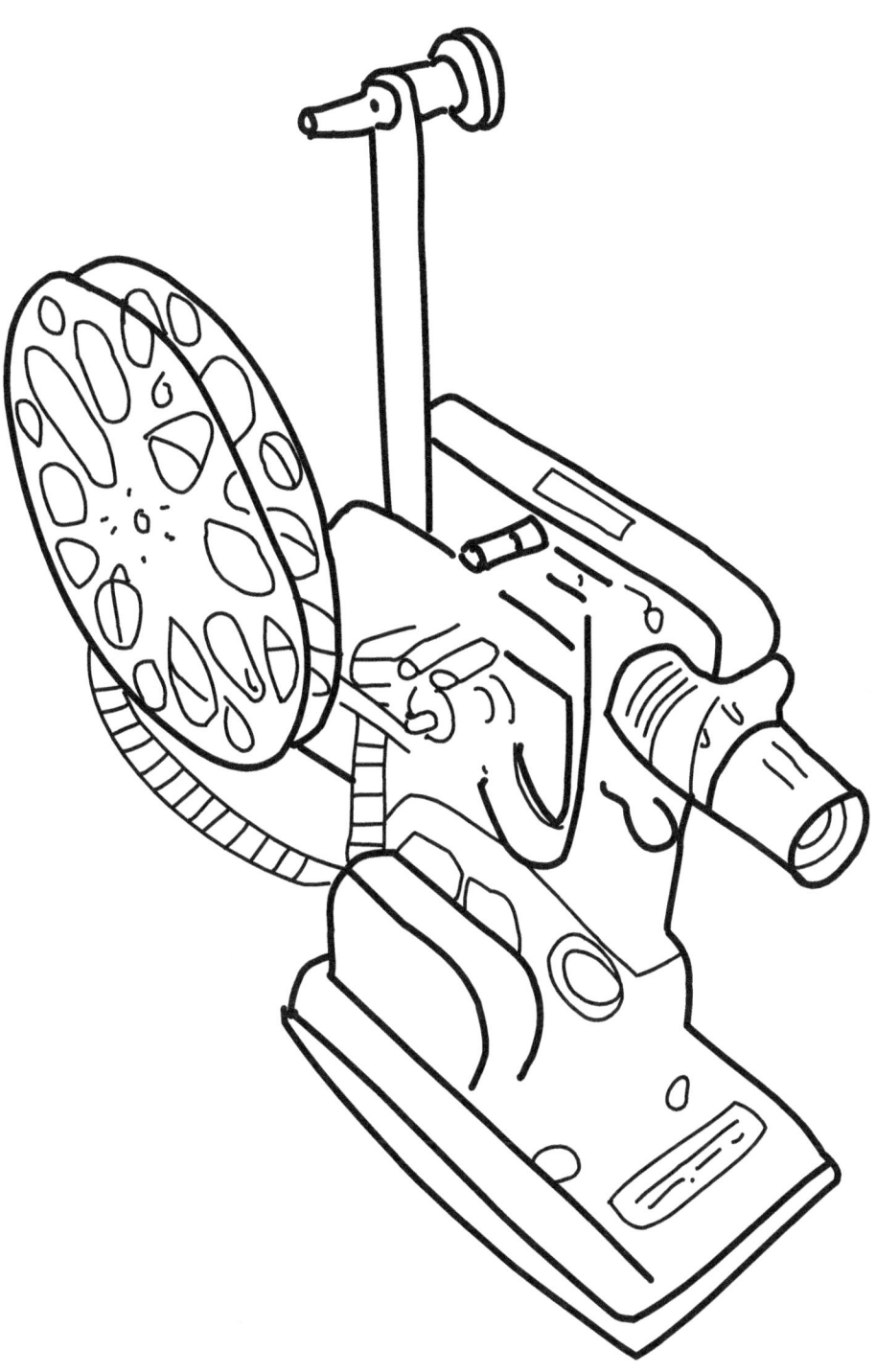

Kinetoscope & Kinetograph

Era
1891–1900s

Aspect Ratio
1.33:1 (4:3)

Developed by
William Kennedy Dickson
(Edison Company)

Format
analog

Size
18mm

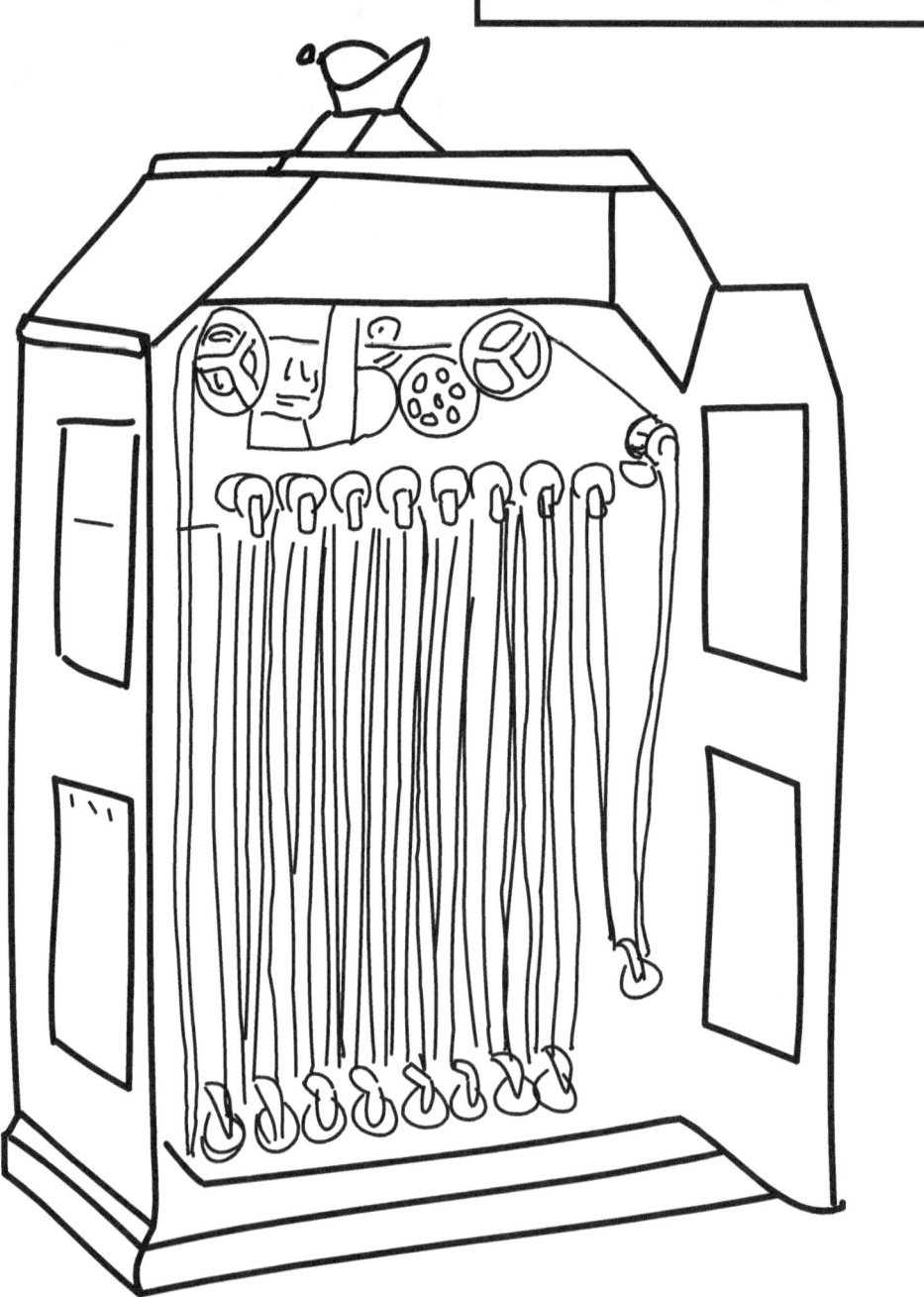

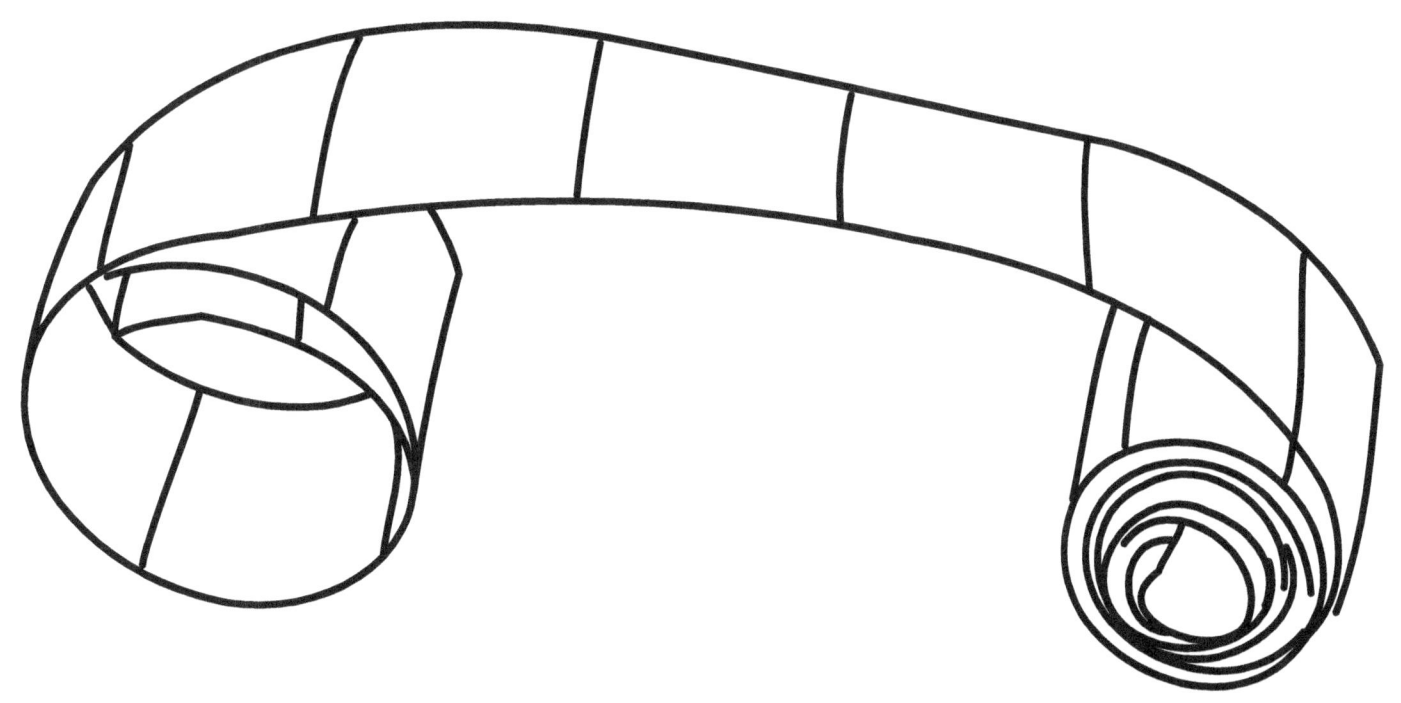

Cinematograph

Era	Size	Aspect Ratio	AKA
1892–1900s	**35mm**	**1.33:1 (4:3)**	**Cinématographe, Kinematograph**

Format	Developed by
analog	**Léon Bouly**

Fun fact
**This machine
weighed 16 lbs
and was
hand-cranked**

Fun fact
**This format
was a projector
and camera in
the same device
(for the first time)**

Fun fact
**Invented by Léon Bouly,
this format was adopted
and popularized by
the Lumière brothers**

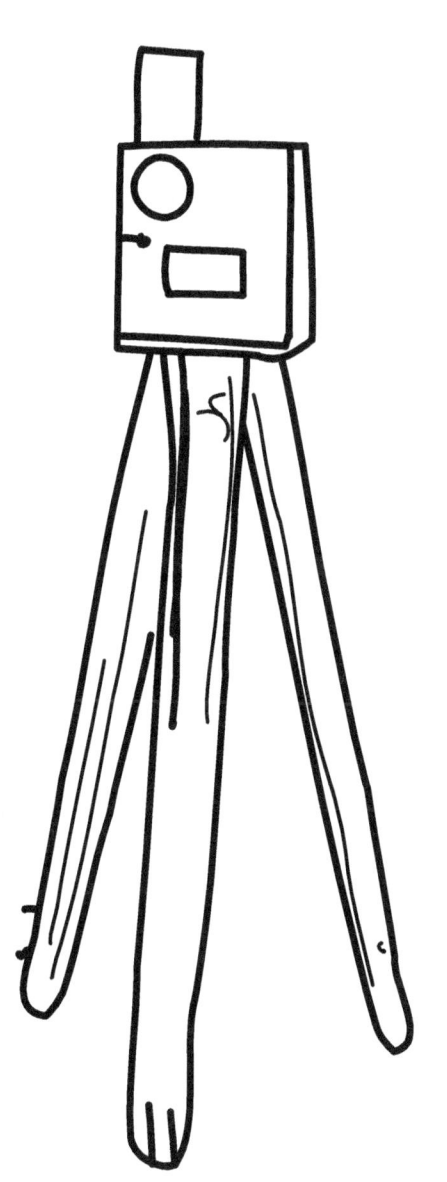

Eidoloscope

Era
1894–1900s

Size
51mm

Aspect Ratio
1.85:1

AKA
**Pantoptikon,
Panoptikon**

Developed by
**Eugene Augustin Lauste
& Woodville Latham
(Lambda Company)**

Fun fact
**This was likely the first
widescreen film format,
with an aspect ratio of 1.85:1**

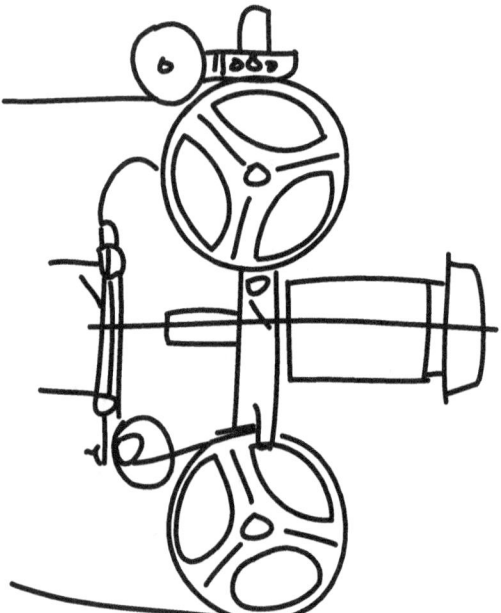

Fun fact
**This was the first
device to use the
Latham loop, a
method that
allowed film
strips to
move quickly
and gently
through a
projector**

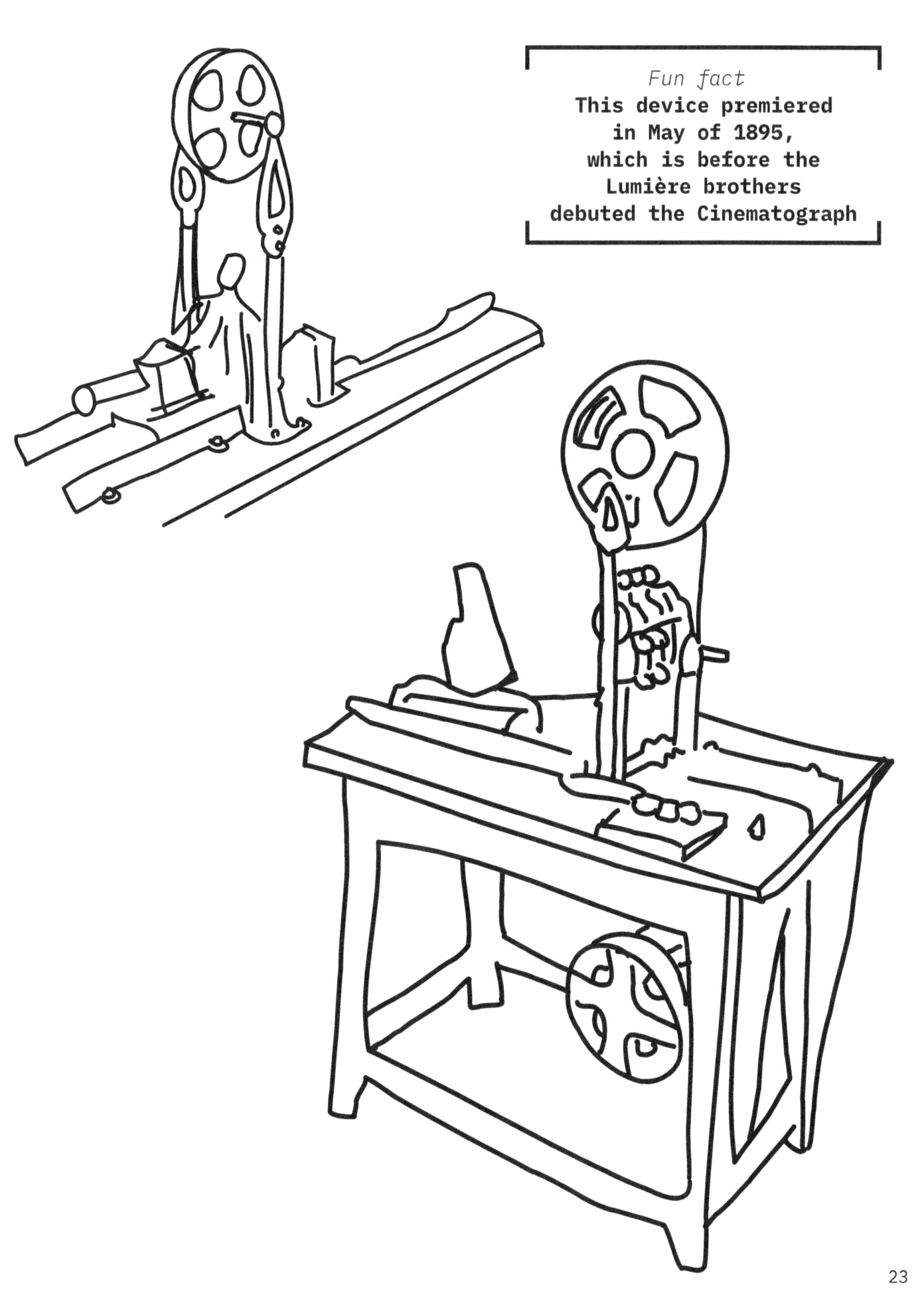

Fun fact
This device premiered
in May of 1895,
which is before the
Lumière brothers
debuted the Cinematograph

23

Bioscop

Era
1895–1900s

Size
54mm

Aspect Ratio
1.33:1 (4:3)

AKA
Bioskop, Bioscope

Developed by
Max Skladanowsky

Fun fact
This format used two loops of 54mm filmstrips with no side perforations during filming and 4 perforations during projection

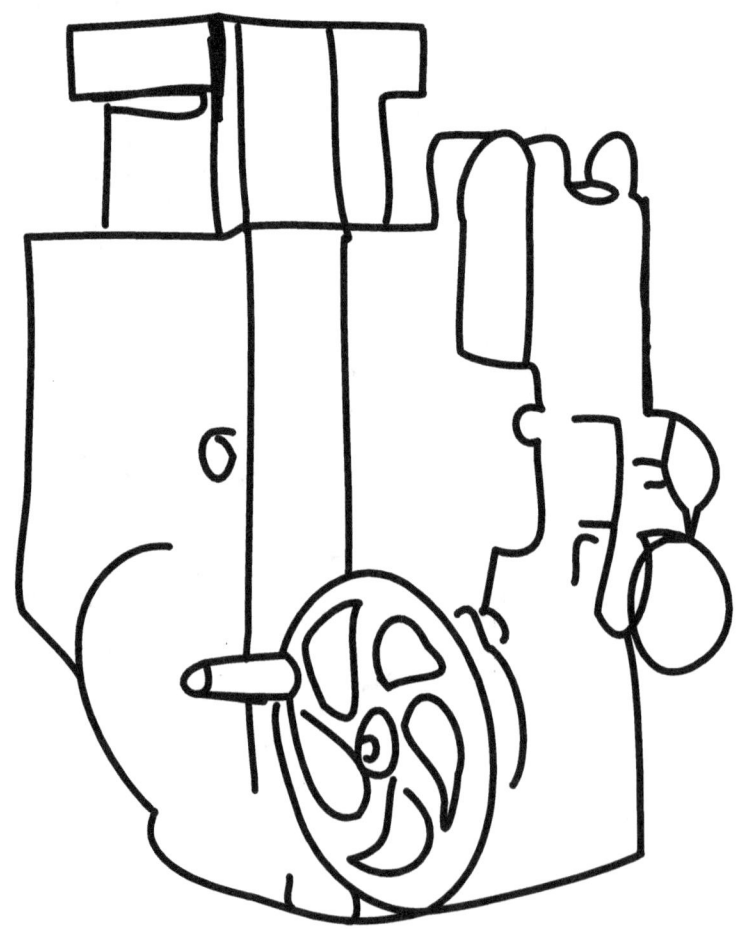

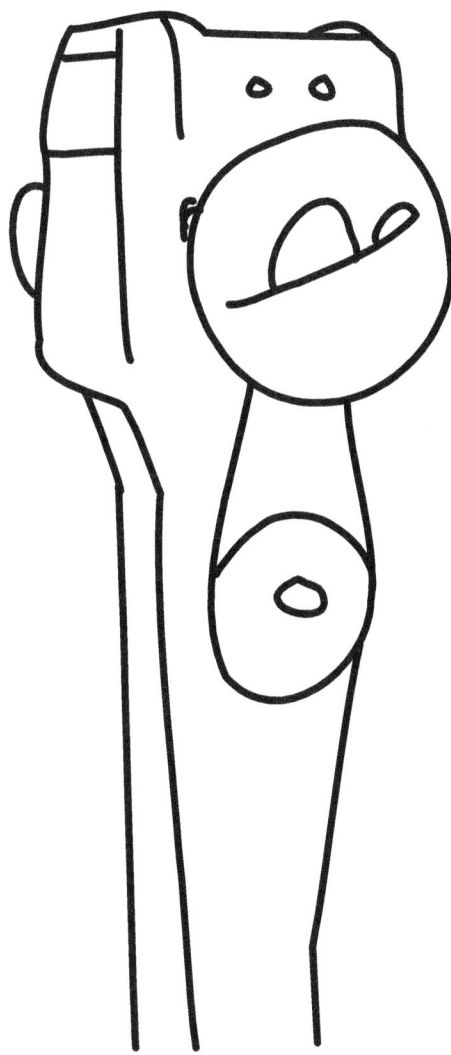

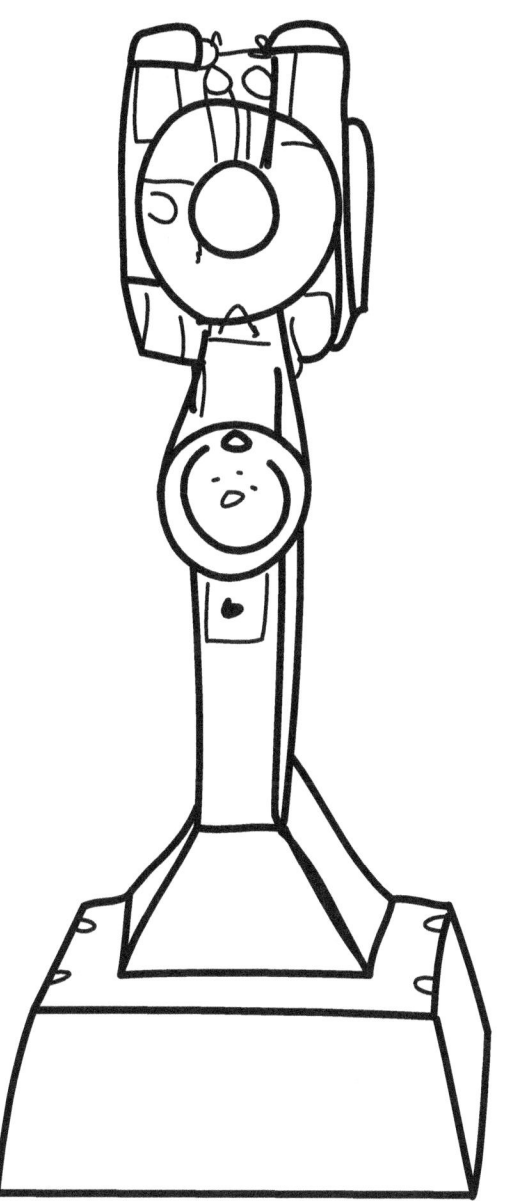

Fun fact
**This system worked
by projecting one
frame from each strip
of film in rapid succession**

Fun fact
**Footage from
this camera
debuted in
July 1895 and
Skladanowsky
was awarded
a patent in
November 1895**

Biograph & Mutograph

Format
analog

AKA
68mm

Era
1897–1902

Developed by
**William Kennedy Dickson
& Herman Casler
(Biograph Company)**

Aspect Ratio
1.35:1

Size
68mm

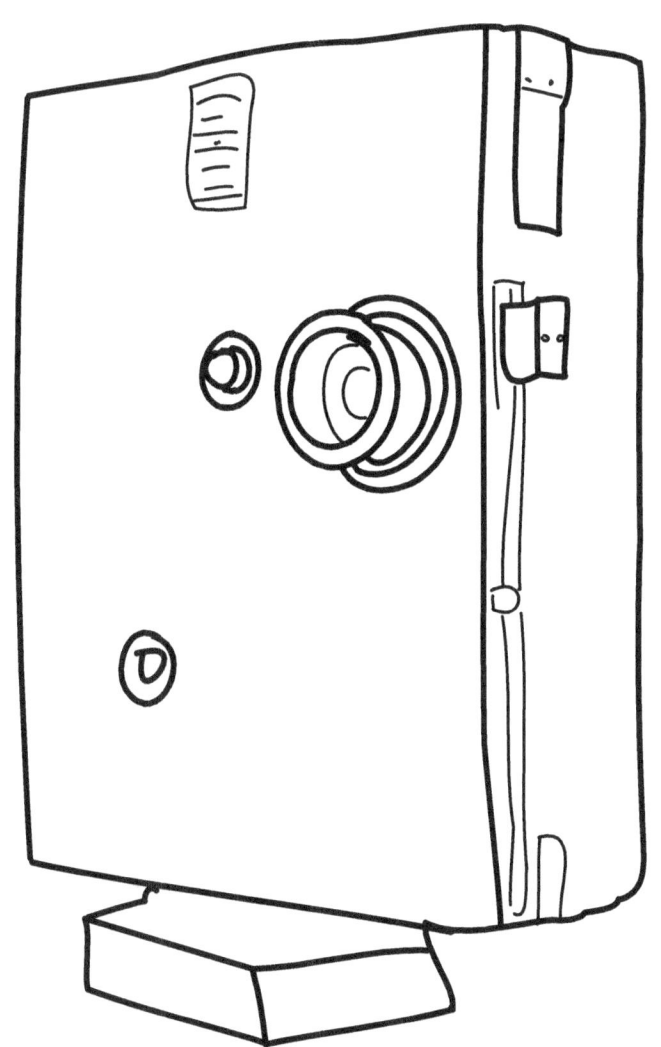

Fun fact
**The Mutograph was
the recording camera
and the Biograph
was the name
of the projector**

Fun fact
**This format had
no sprocket holes,
unless they were
added to help
projection**

Fun fact
**Early versions
of this camera
used a larger
format to avoid
Edison's early
motion picture
patents**

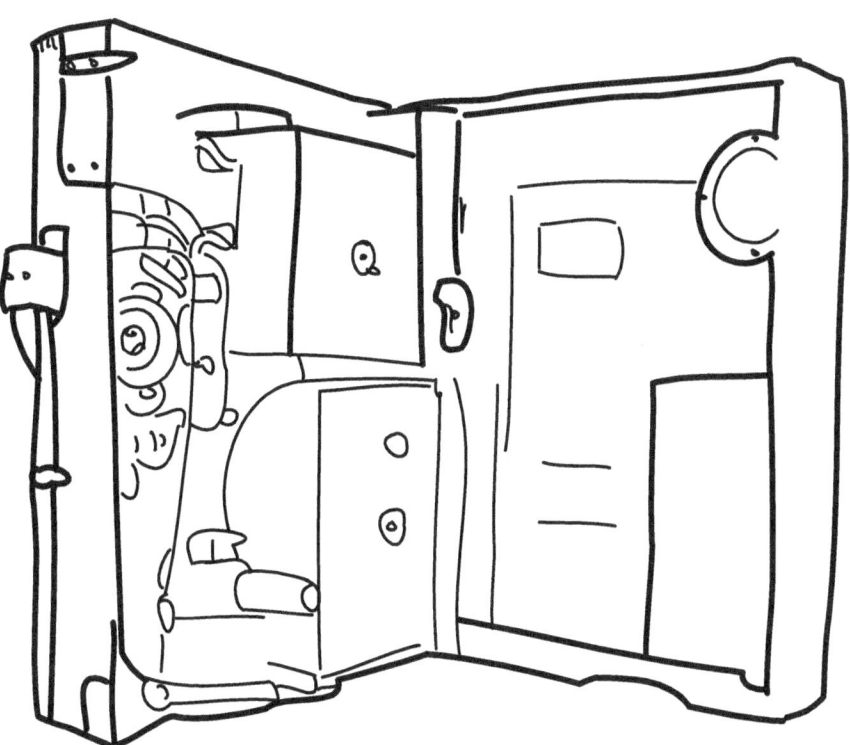

Cinéorama

Era
1897

Size
70mm

Aspect Ratio
1 (360°)

Developed by
Raoul Grimoin-Sanson

Fun fact
**The projection
was successful
but the projection
room would
heat up to the
point that on
one occasion
the operator
fainted, making
it too dangerous
to use**

Fun fact
This system
premiered at
the 1900 Paris
World's Expo

Fun fact
**This format
consisted of
10 projectors
projecting on
9x9 meter screens**

17.5mm

Era
1898–1900s

Size
17.5mm

Developed by
Various

Aspect Ratio
Various

Fun fact
This format could be created by cutting strips of 35mm film in half

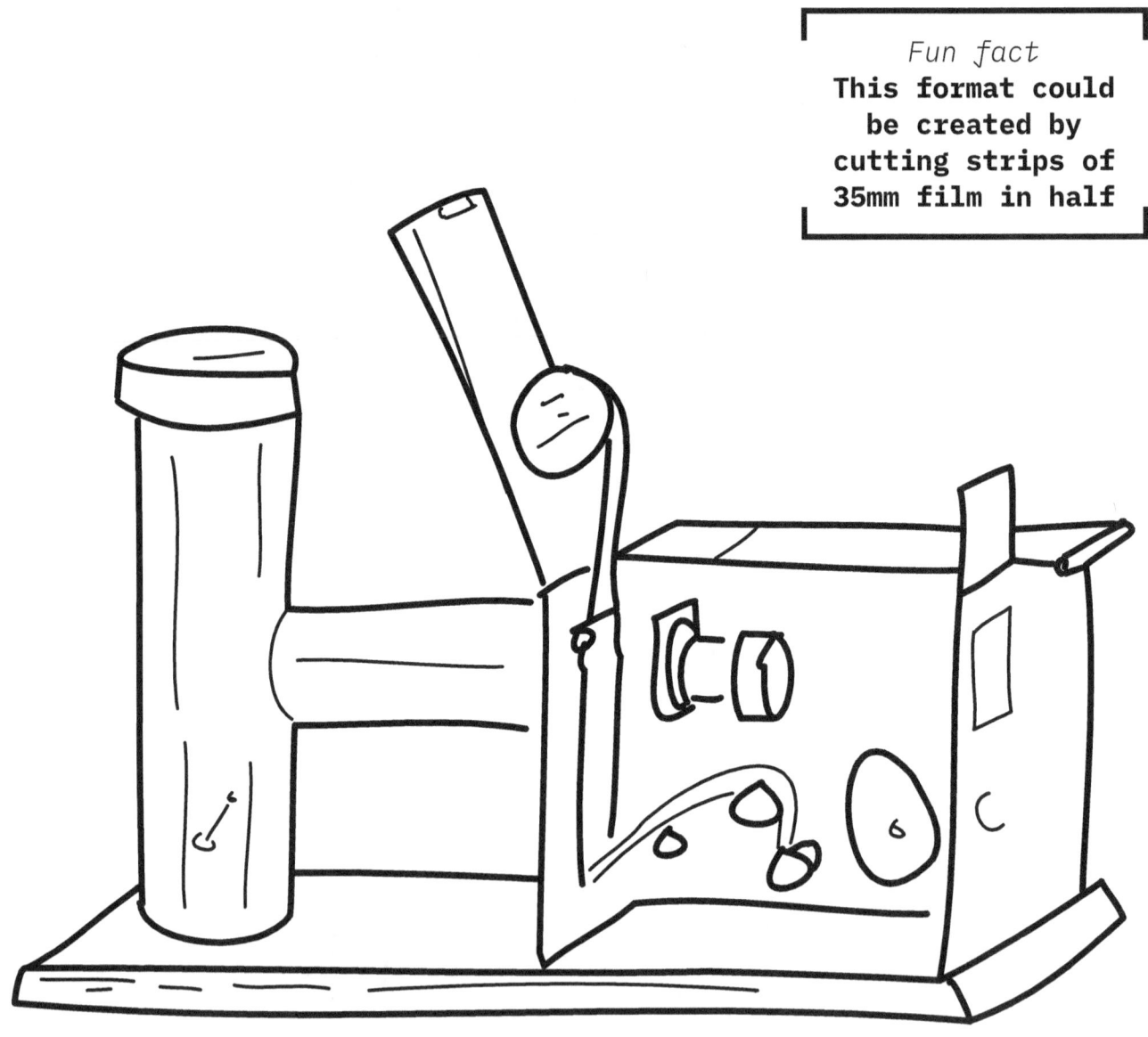

Fun fact
Several formats were developed to use this size, eight different kinds of film stock: Birtac, Biokam, Hughes, Gaumont, Clou, Duoscope, Movette, and Pathé Rural

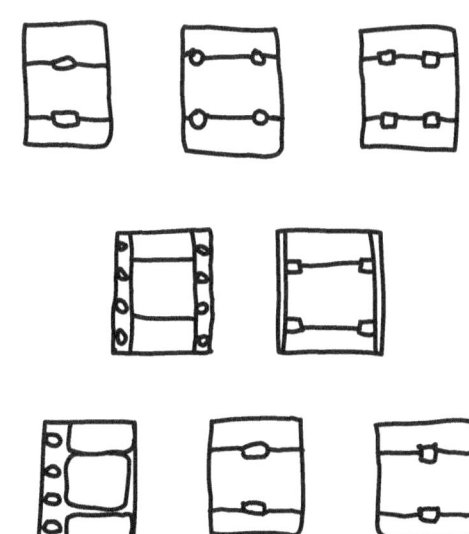

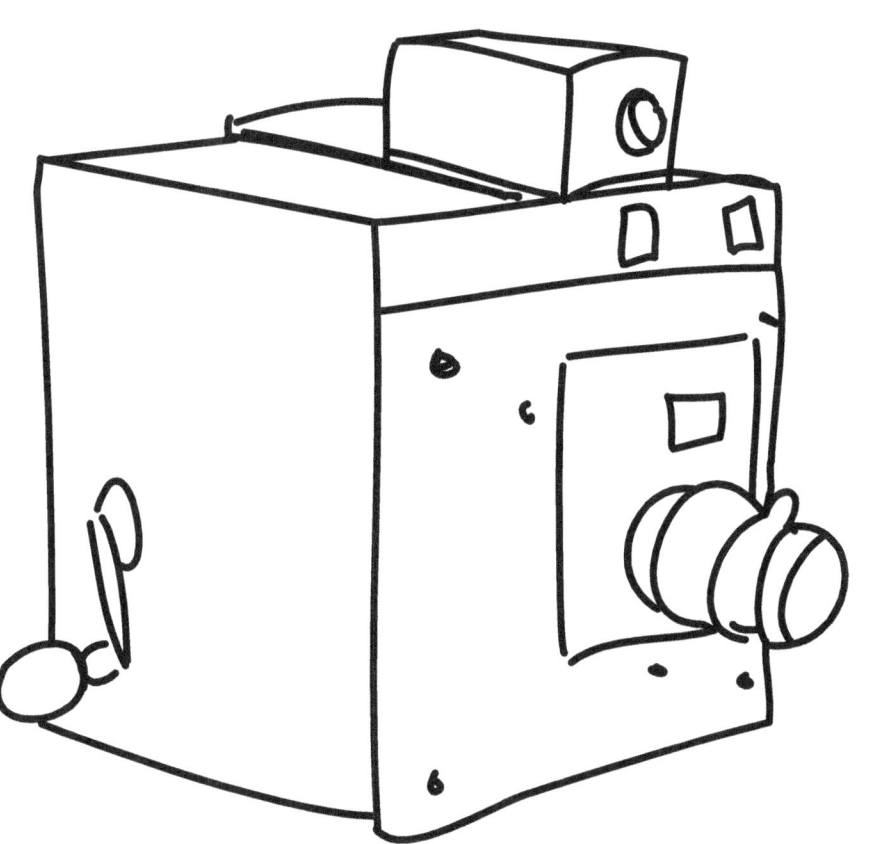

Fun fact
The first use of this format was with the Birtac, which was a combined projector and camera

Home Kinetoscope

Format
analog

Era
1911–1915

Size
22mm

Aspect Ratio
1.33:1 (4:3)

AKA
**Projectoscope,
Projecting Kinoscope**

Developed by
Edison Company

Fun fact
**This format
was a
projecting
device
and a
dedicated
camera
was never
created**

Fun fact
**This format
consisted of
three rows of
images separated
by two rows
of perforations**

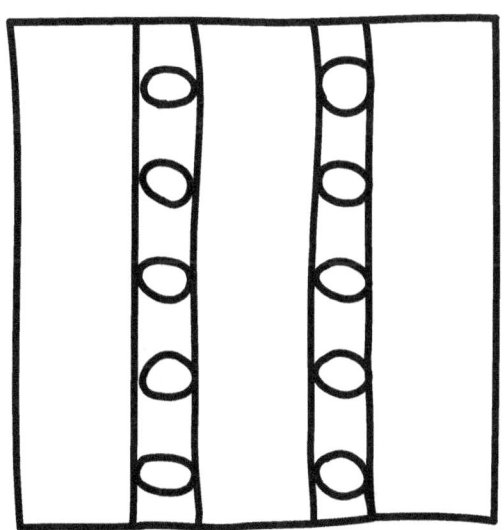

Fun fact
**For the three film
strips, a column
of images was
cranked forward,
the middle row
backward, and
the third row
forward again**

28mm

Format
analog

Size
28mm

Era
1912–1920

Developed by
Pathé

Aspect Ratio
1.36:1

AKA
Pathé Kok

Fun fact
**This format struggled to compete in the
market against 9.5mm and 16mm film**

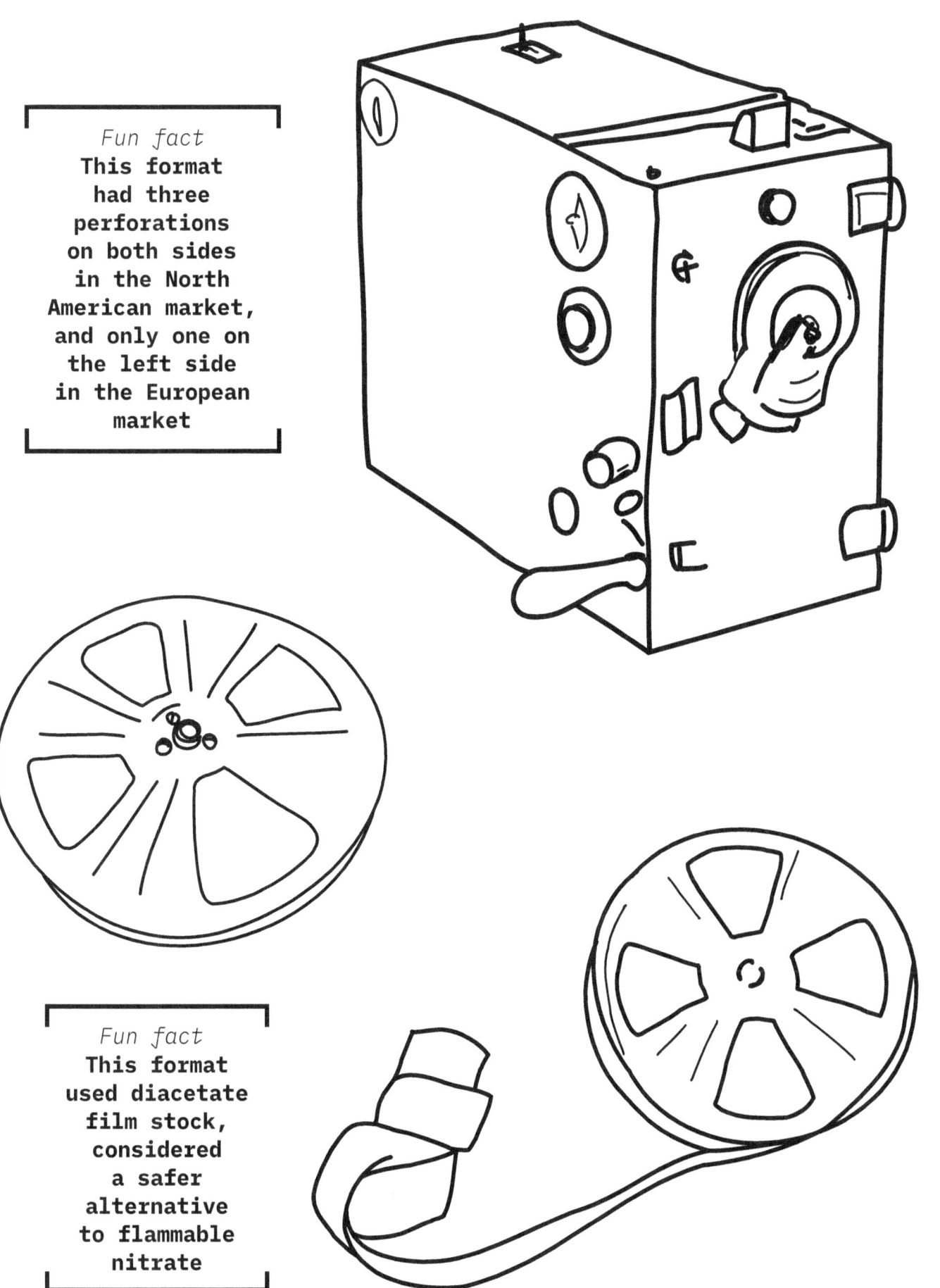

Fun fact
**This format
had three
perforations
on both sides
in the North
American market,
and only one on
the left side
in the European
market**

Fun fact
**This format
used diacetate
film stock,
considered
a safer
alternative
to flammable
nitrate**

9.5mm

Format
analog

Size
9.5mm

Era
1922–1960

Developed by
Pathé

Aspect Ratio
1.31:1

AKA
Pathé Baby

Fun fact
This format was mostly popular in European markets

Fun fact
This format had a single sprocket hole in the middle of the filmstrip, between frames

Fun fact
This format was used for commercial films and home movies

16mm

Format	Size	Era
analog	**16mm**	**1923–**

AKA	Developed by	Aspect Ratio
Safety film	**Kodak**	**1.37:1**

Fun fact
Originally silent, this format was fitted to include recorded sound in 1932 by RCA

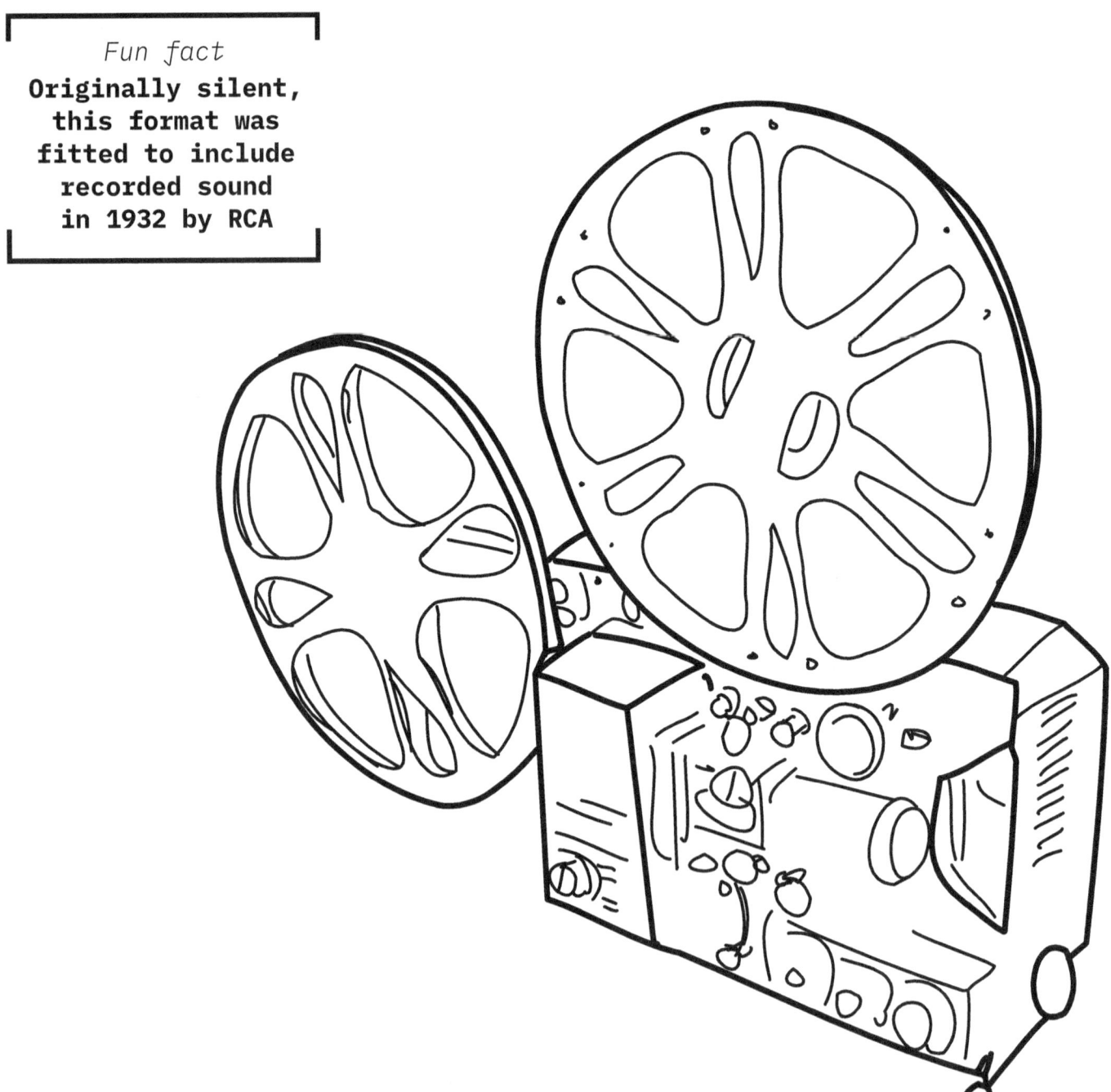

Movietone

Size
35mm

Aspect Ratio
1.16:1

Era
1926–1930s

Developed by
Fox

Fun fact
This format was used for newsreels and a handful of motion pictures

Fun fact
This format had its own specific aspect ratio to accomodate the additional printed soundtrack; other formats used Silent (1.33:1) or Academy (1.375:1) aspect ratios

Fun fact
This was one of four sound technologies in development during the 1920s; the others were DeForest Phonofilm, Warner Brothers' Vitaphone, and RCA Photophone

Polyvision

Developed by
Abel Gance

Era
1927

Size
35mm × 3

Aspect Ratio
4:1

Fun fact
**This format had a three-camera rig
which allowed three cameras
to shoot a single scene in panorama**

Fun fact
This format was only used to produce a single film, Abel Gance's *Napoleon* (1927)

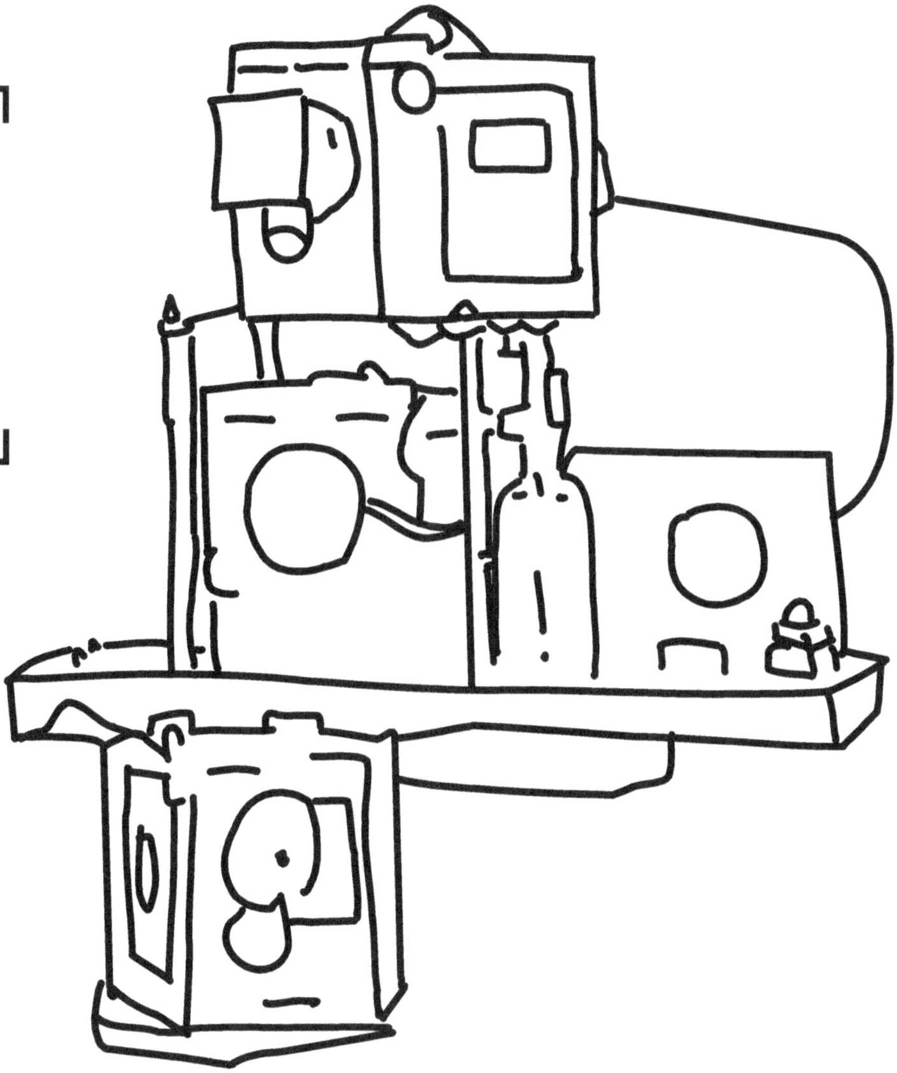

Fun fact
This format was recorded with three cameras vertically stacked, and projected with three projectors, horizontally arranged

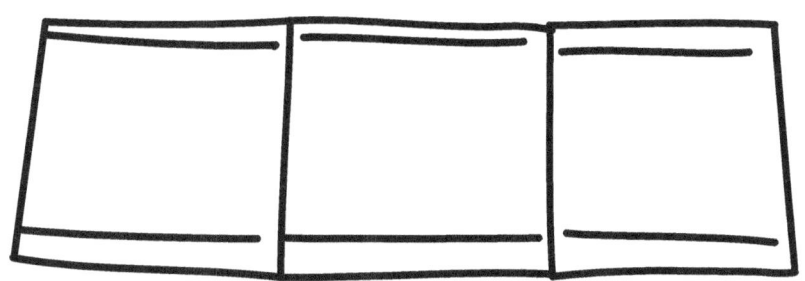

Fox Grandeur

Format
analog

Era
1928–1931

Developed by
Fox Film Corporation

Size
70mm

Aspect Ratio
2:1

AKA
Grandeur 70,
70mm Grandeur film

Fun fact
This format was successful up until the advent of the Great Depression, whereafter it failed to receive adequate funding

Fun fact
This format used Movietone sound on film technology

Fun fact
**This format didn't use
the standard 35mm
perforation size;
like the film itself, the
perforations were larger**

35mm (Sound)

Format
analog/digital

Aspect Ratio
1.375:1

Developed by
William Kennedy Dickson
& Thomas Edison
(Edison Company)

Era
1929–

Size
35mm

AKA
Academy

Fun fact
This process can include either an analog or digital soundtrack, and the signal can be recorded optically or magnetically

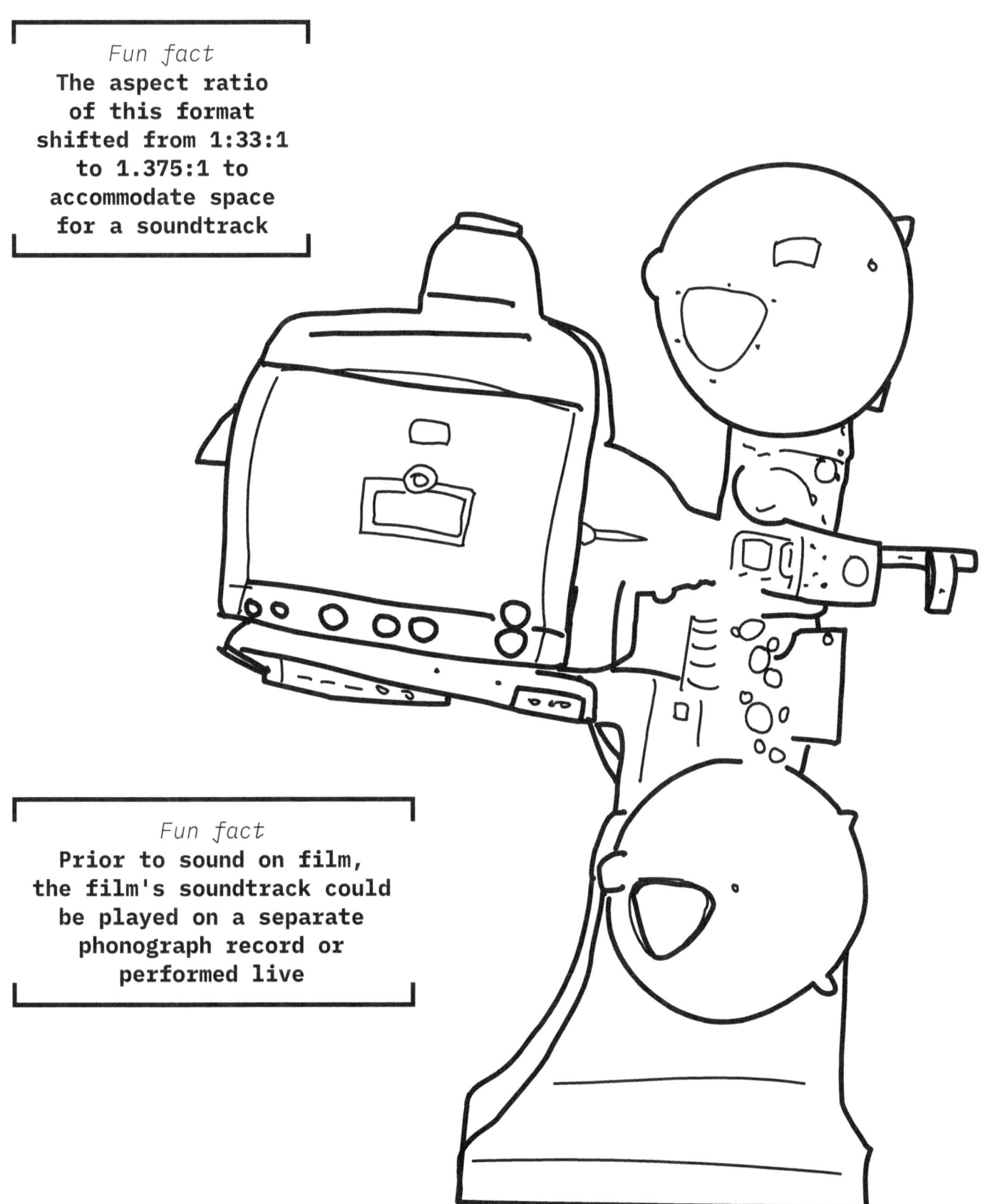

Hypergonar

Developed by
Henri Chrétien

Size
35mm

Aspect Ratio
2.66:1

Era
1929-

AKA
Anamorphoscope

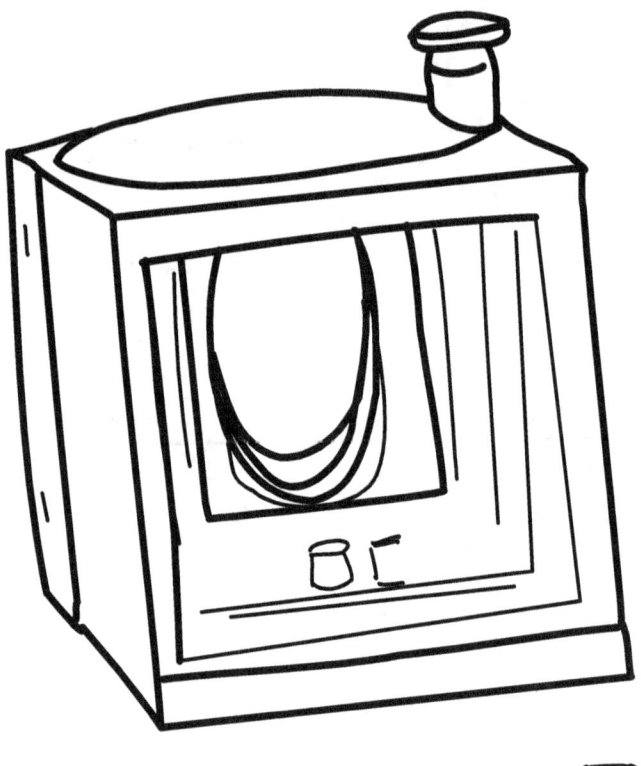

Fun fact
This lens technology influenced many anamorphic formats popular in the 1950s

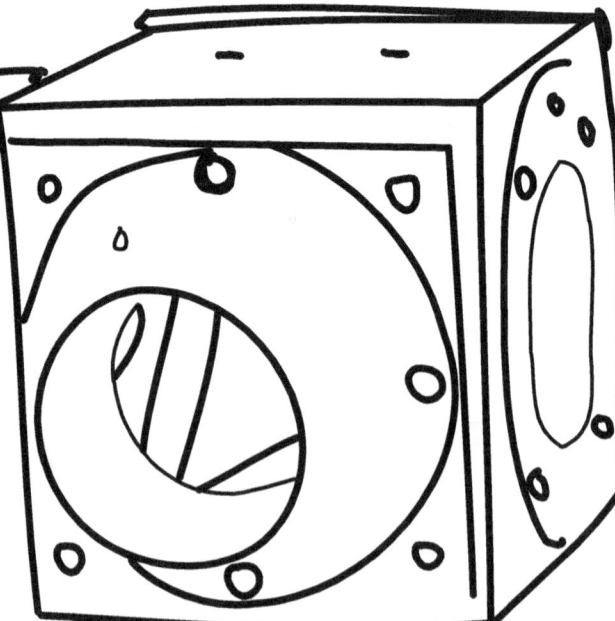

8mm

Format
analog

Size
8mm

Developed by
Eastman Kodak

Era
1932–

Aspect Ratio
1.32:1

Fun fact
This format was created to be a smaller, cheaper alternative to 16mm film

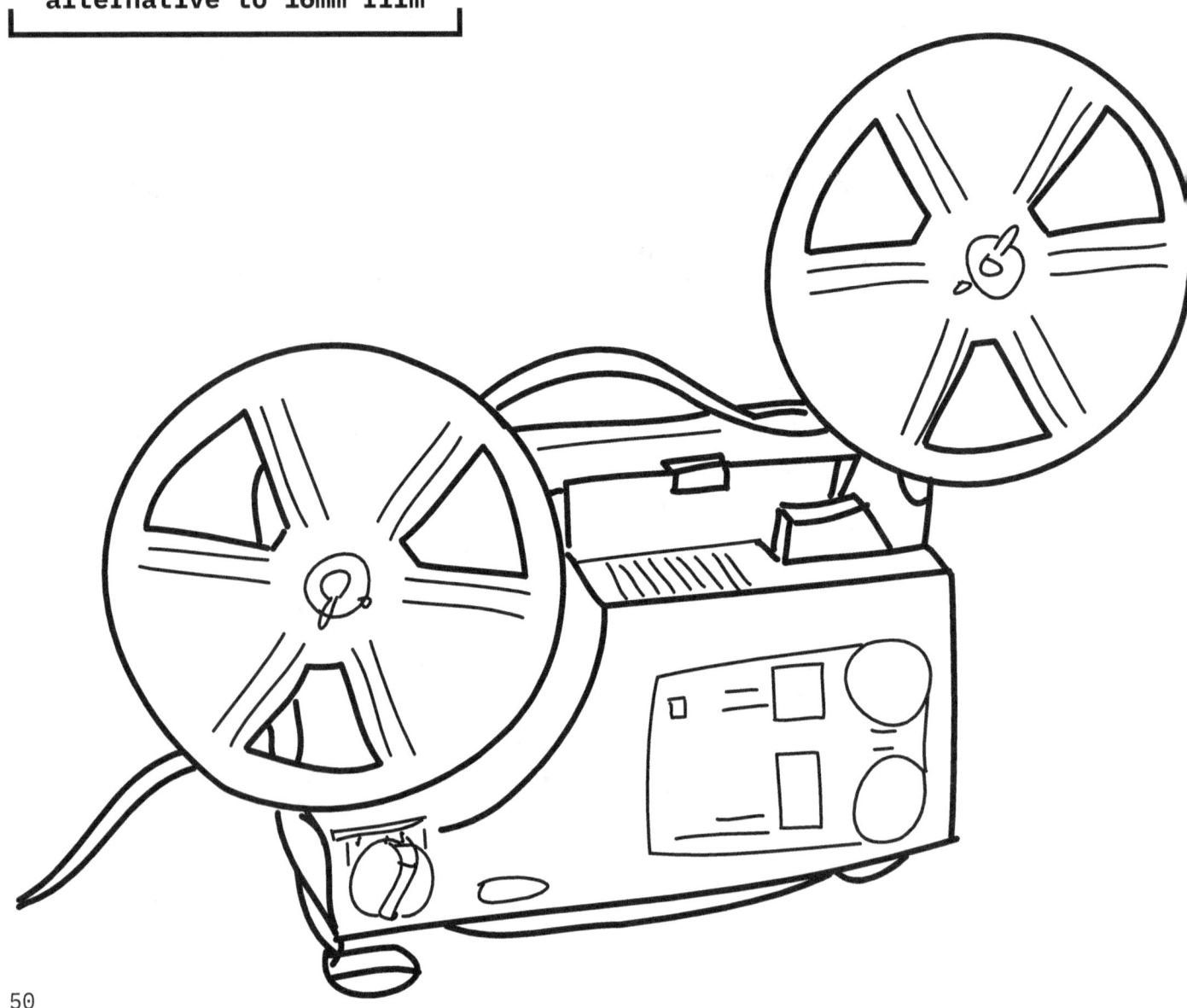

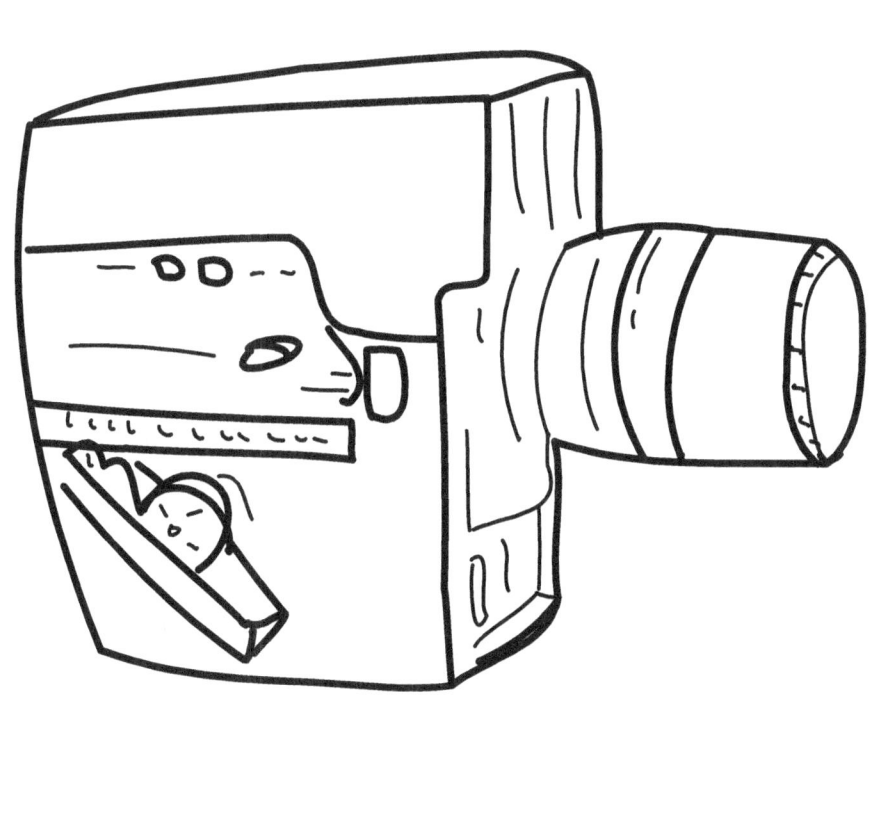

Fun fact
**This format was
designed to
be silent but
the ability to
add sound was
introduced in
the 1960s**

Fun fact
**Film in this format
ran at 12, 15, 16,
or 18 frames per second**

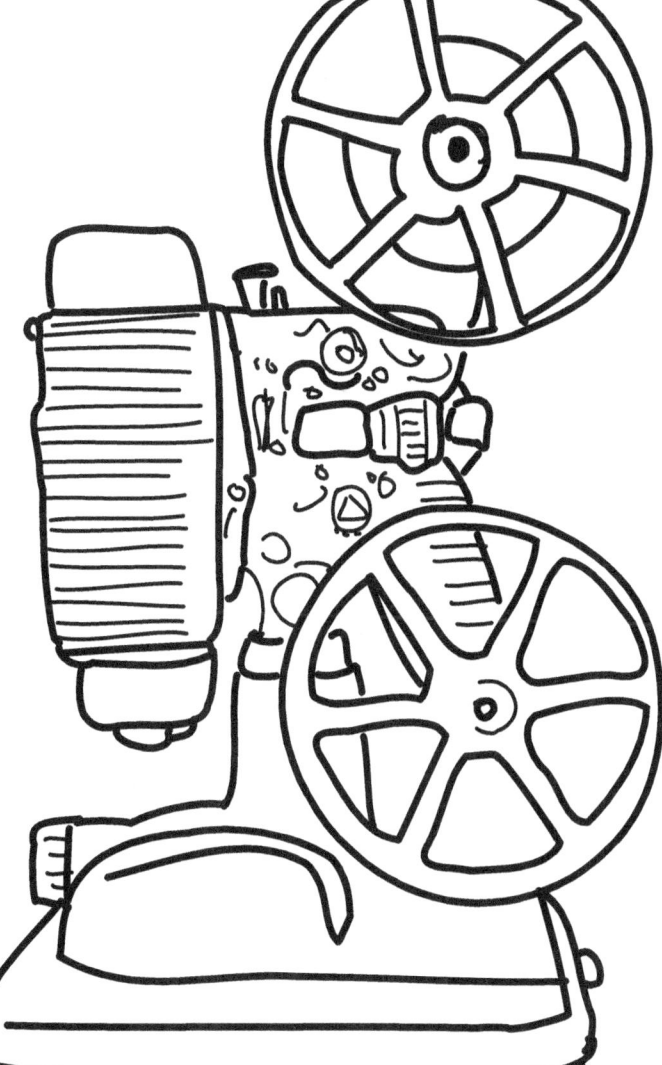

Cinerama

Developed by
Cinerama Corporation

Aspect Ratio
2.59:1

Fun fact
This format released a 70mm widescreen edition, which was not successful

Era
1952-2020

Size
35mm × 3

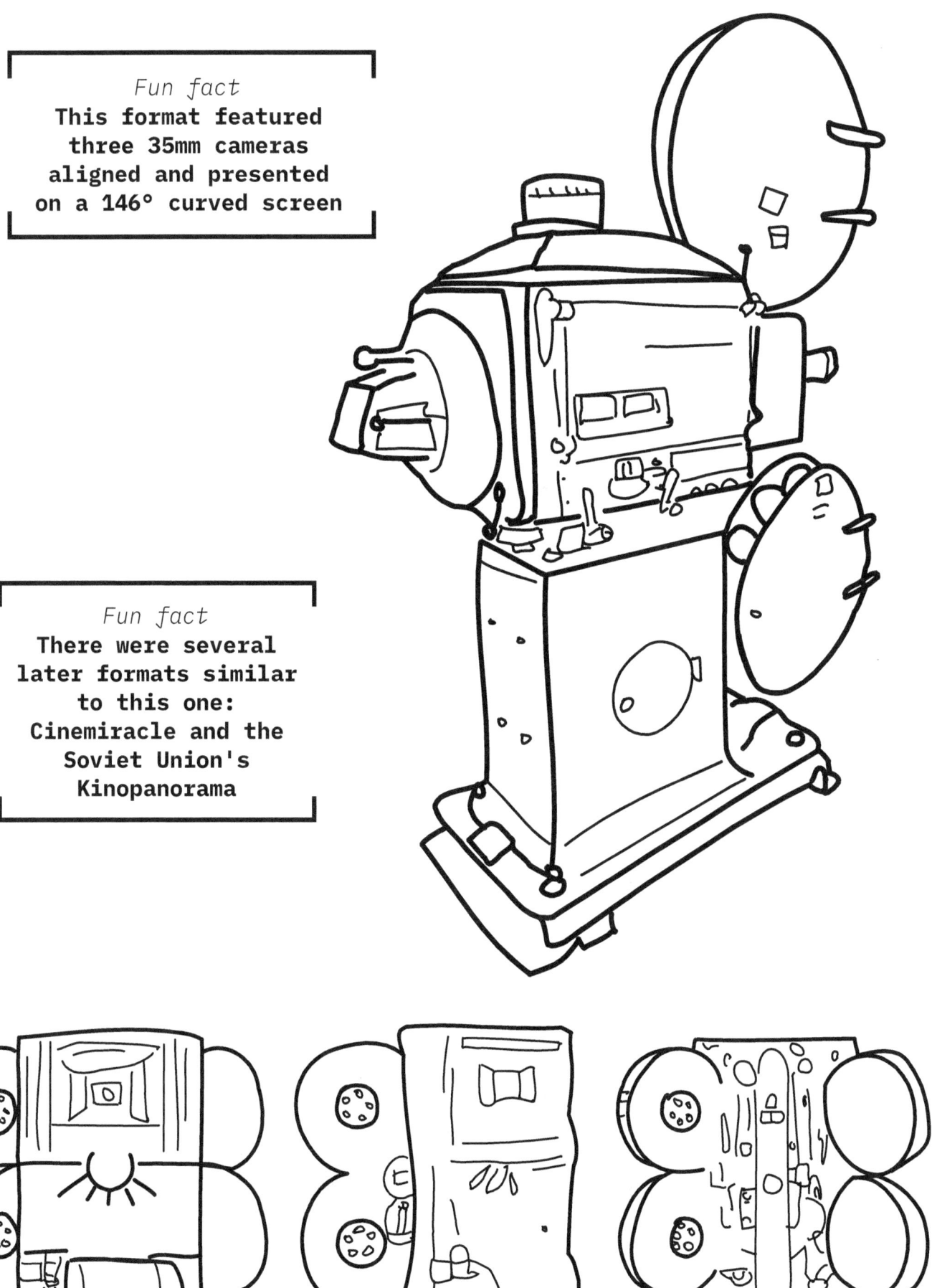

CinemaScope

54

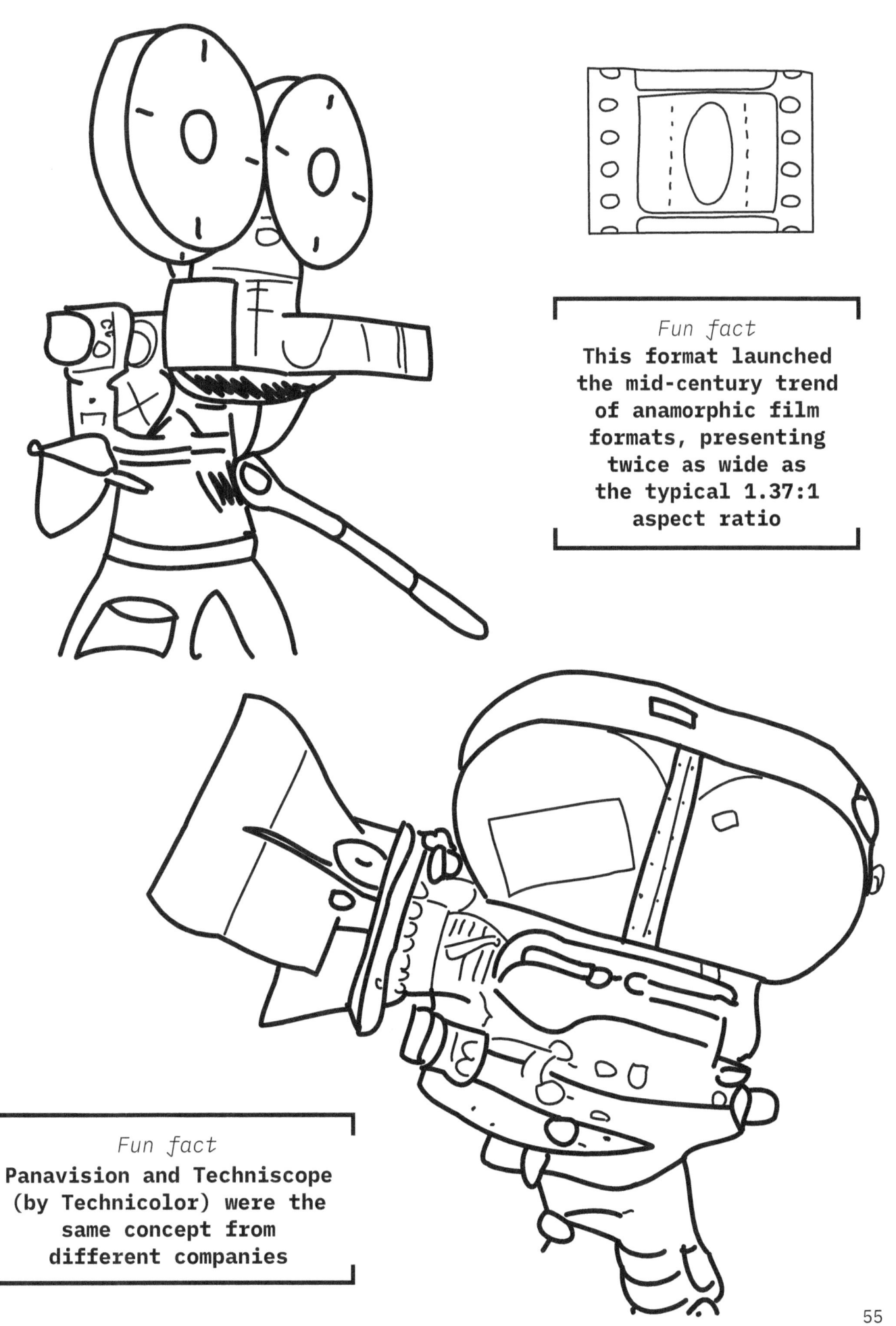

VistaVision

Developed by
Paramount

Era
1954-1961

Size
35mm

Aspect Ratio
1.5:1

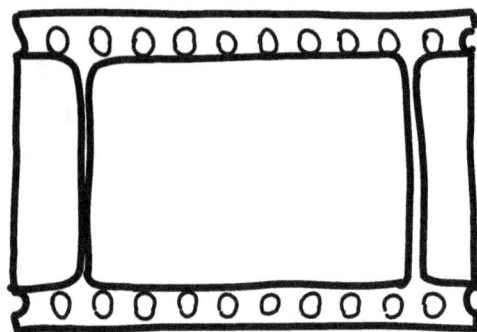

Fun fact
This format had a natural aspect ratio of 1.5:1 but could be further cropped and set to a range of ratios, with motion pictures set in 1.66:1, 1.85:1, and 1.96:1

Fun fact
This format's frames were captured on the film reel horizontally, instead of the traditional vertical

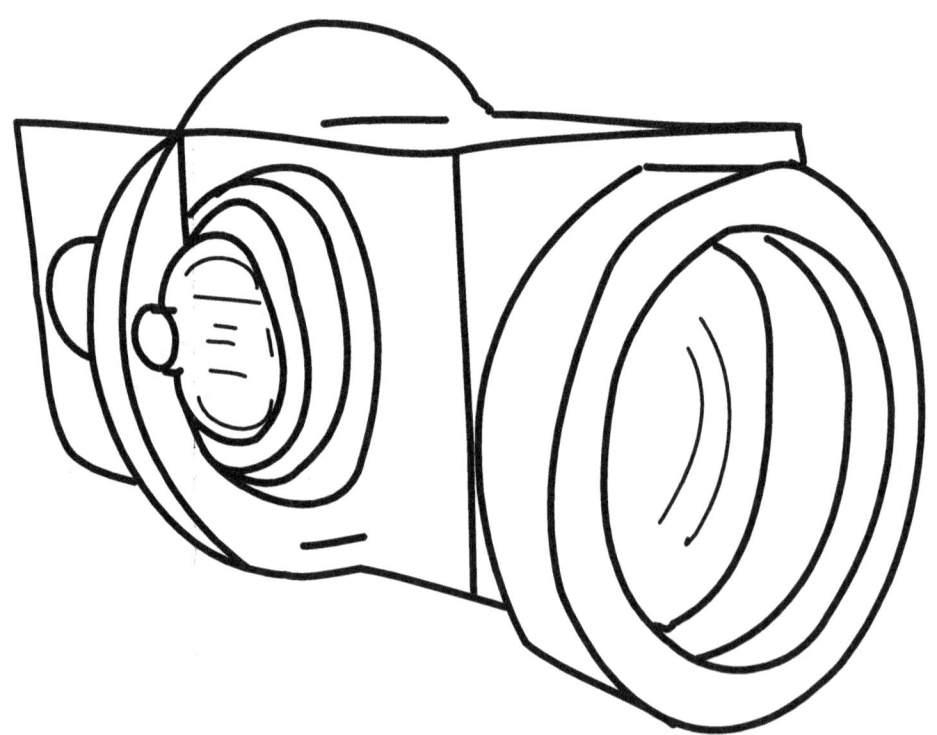

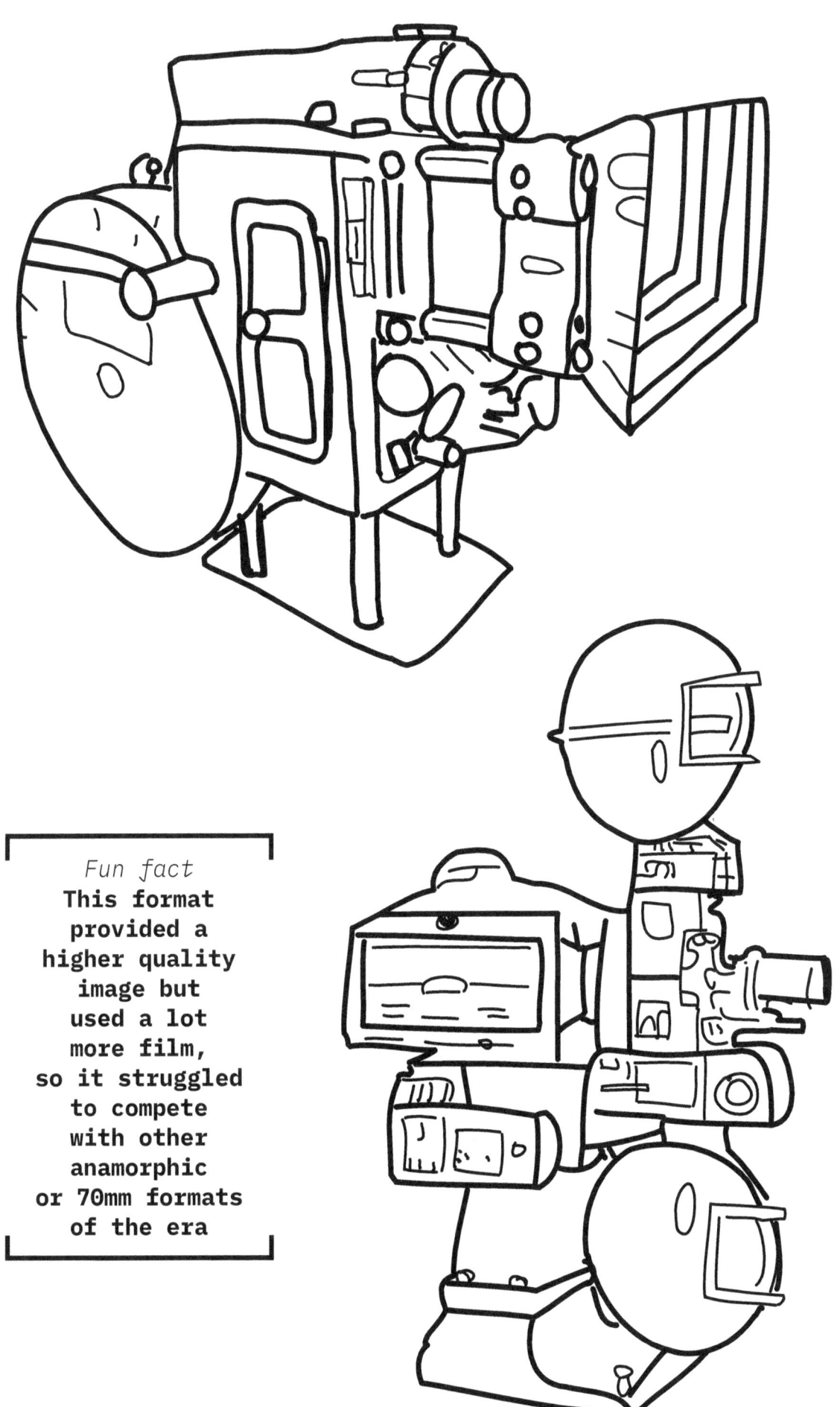

Fun fact
**This format
provided a
higher quality
image but
used a lot
more film,
so it struggled
to compete
with other
anamorphic
or 70mm formats
of the era**

Circle-Vision 360°

Format
analog

Developed by
Disney

Size
35mm × 9

Aspect Ratio
1 (360°)

Fun fact
**Disney had developed
a precursor to
this format called
Circarama, which used
eleven 16mm cameras
to create a circle
of vision**

Era
1955–

Fun fact
**This format was nine
35mm cameras and
fit a projection
screen that wrapped
a full 360°
around the cinema**

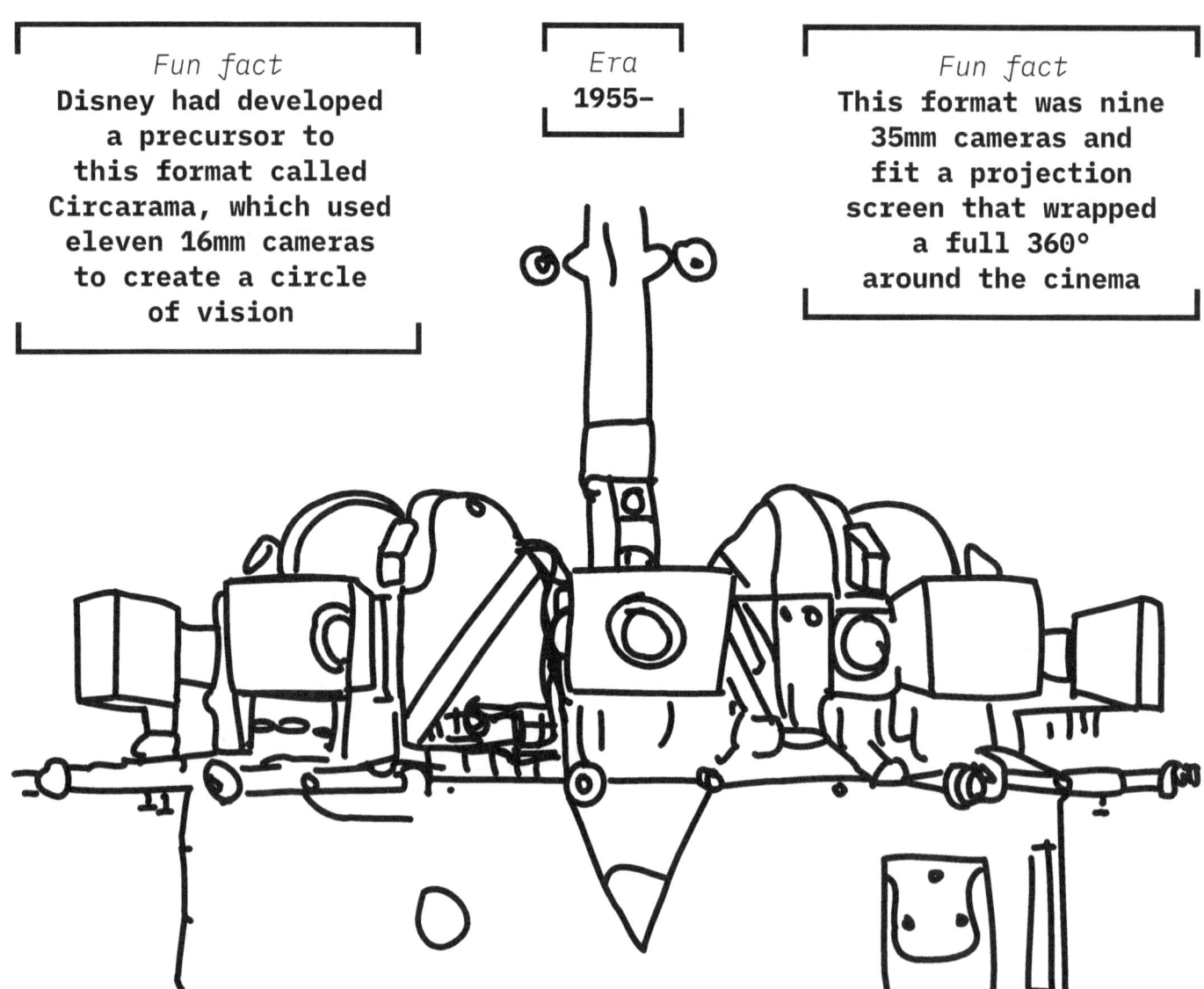

Fun fact
This format premiered at Disneyland on July 17, 1955

Todd-AO

Size
70mm

Era
1955-1970s

Format
analog

Aspect Ratio
2.2:1

Fun fact
This format, created to compete with Cinerama and CinemaScope, was a combination of 70mm film stock mixed with a curved screen

Developed by
Mike Todd, et al

Fun fact
This format had six printed audio tracks

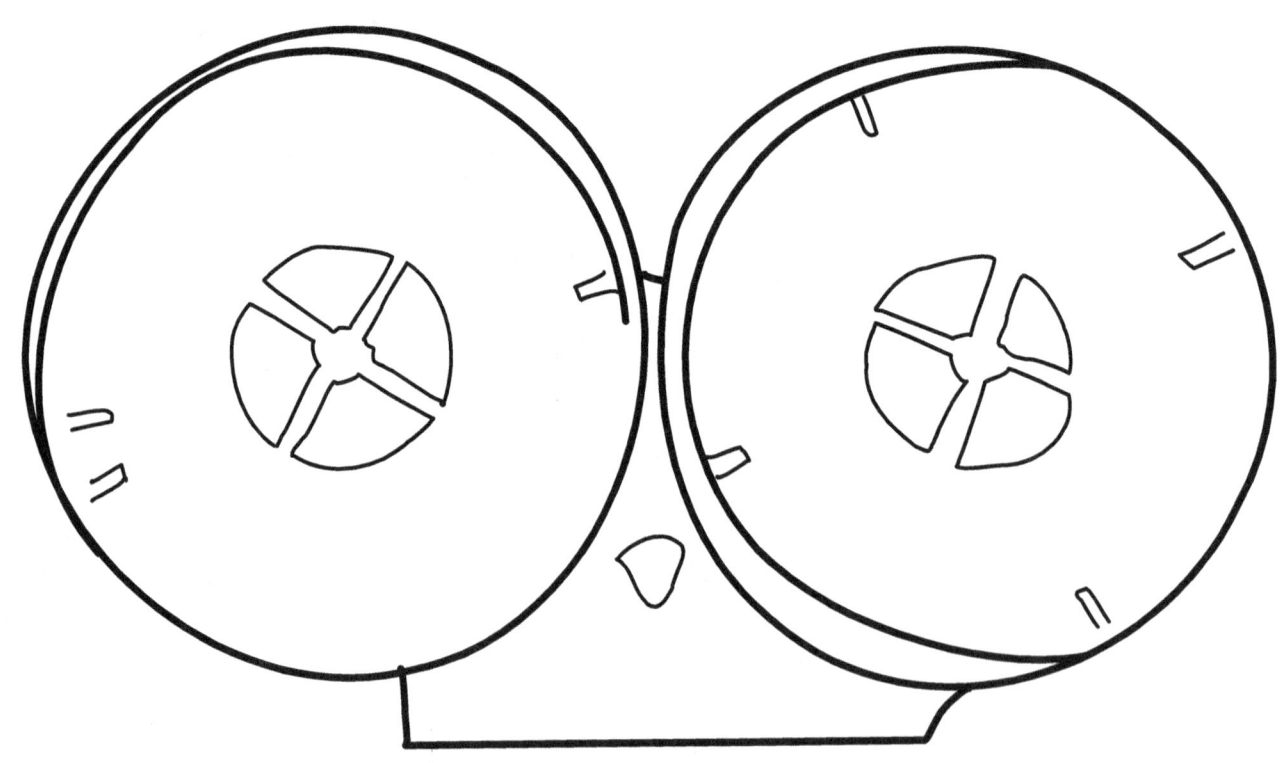

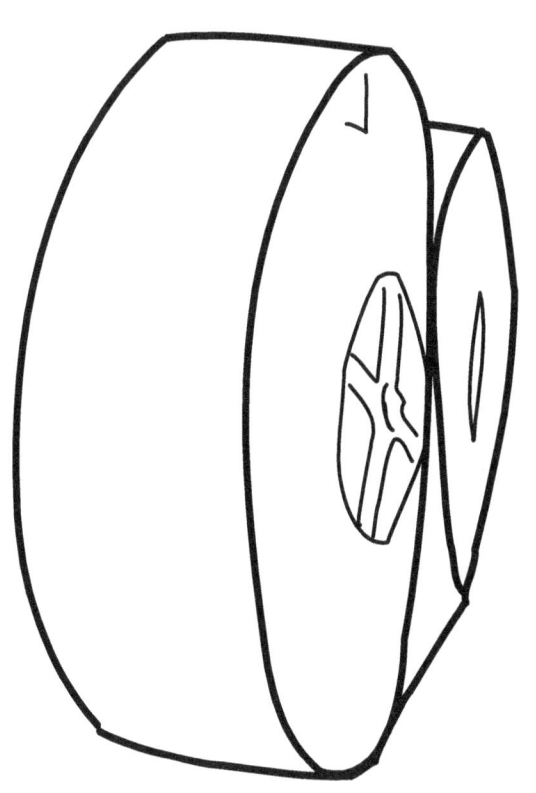

Technirama

Super 8

Size
8mm

Era
1965–

Aspect Ratio
1.36:1

Developed by
Eastman Kodak

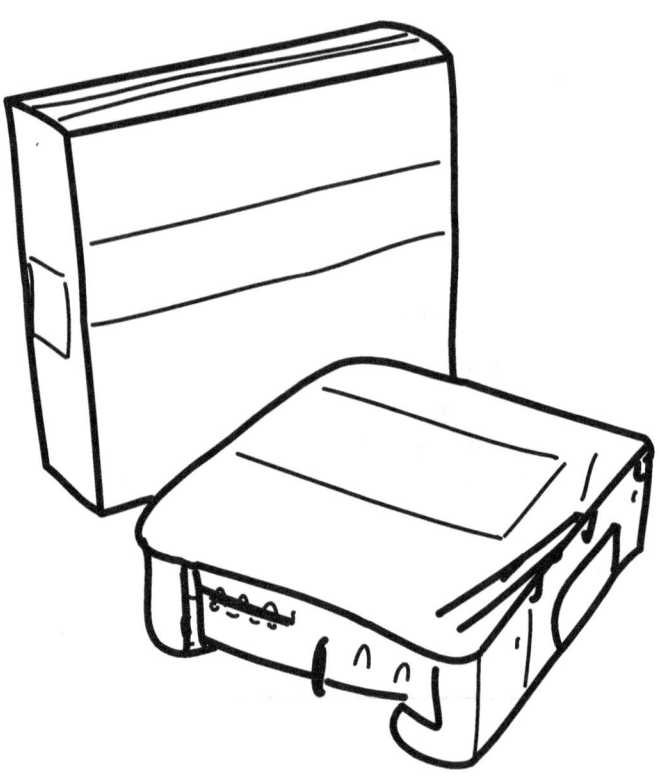

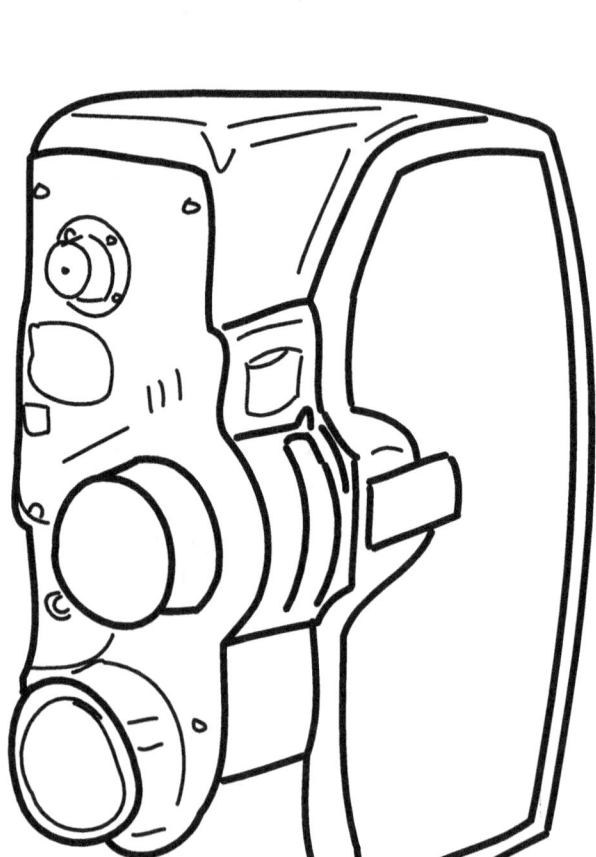

Fun fact
This format is different from its parent format, 8mm, by having smaller perforation holes

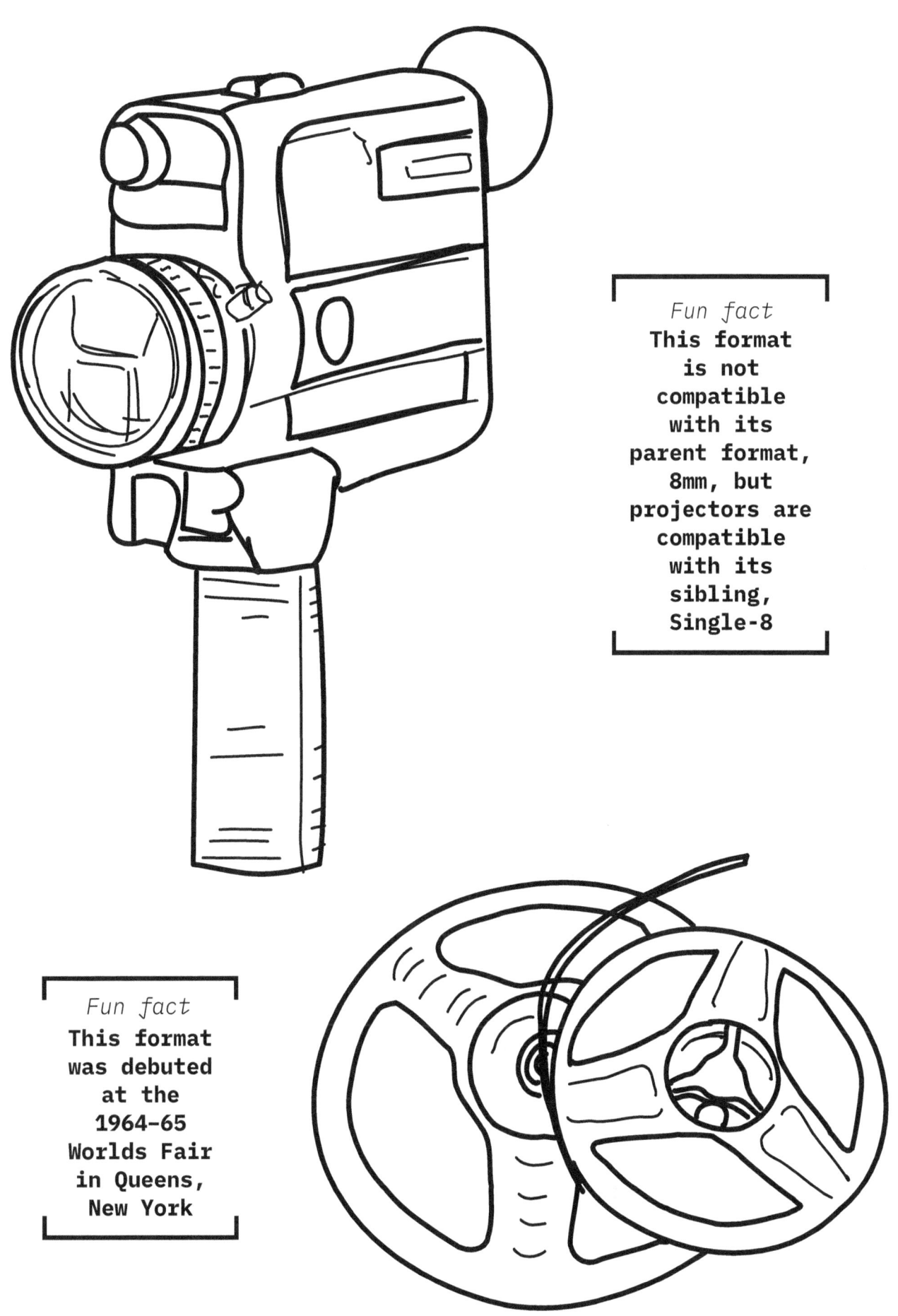

Fun fact
**This format
is not
compatible
with its
parent format,
8mm, but
projectors are
compatible
with its
sibling,
Single-8**

Fun fact
**This format
was debuted
at the
1964-65
Worlds Fair
in Queens,
New York**

Single-8

Developed by
Fujifilm

Era
1965–2012

Size
8mm

Aspect Ratio
1.35:1

Fun fact
**This format was created
by Fujifilm to compete
with Kodak's Super 8**

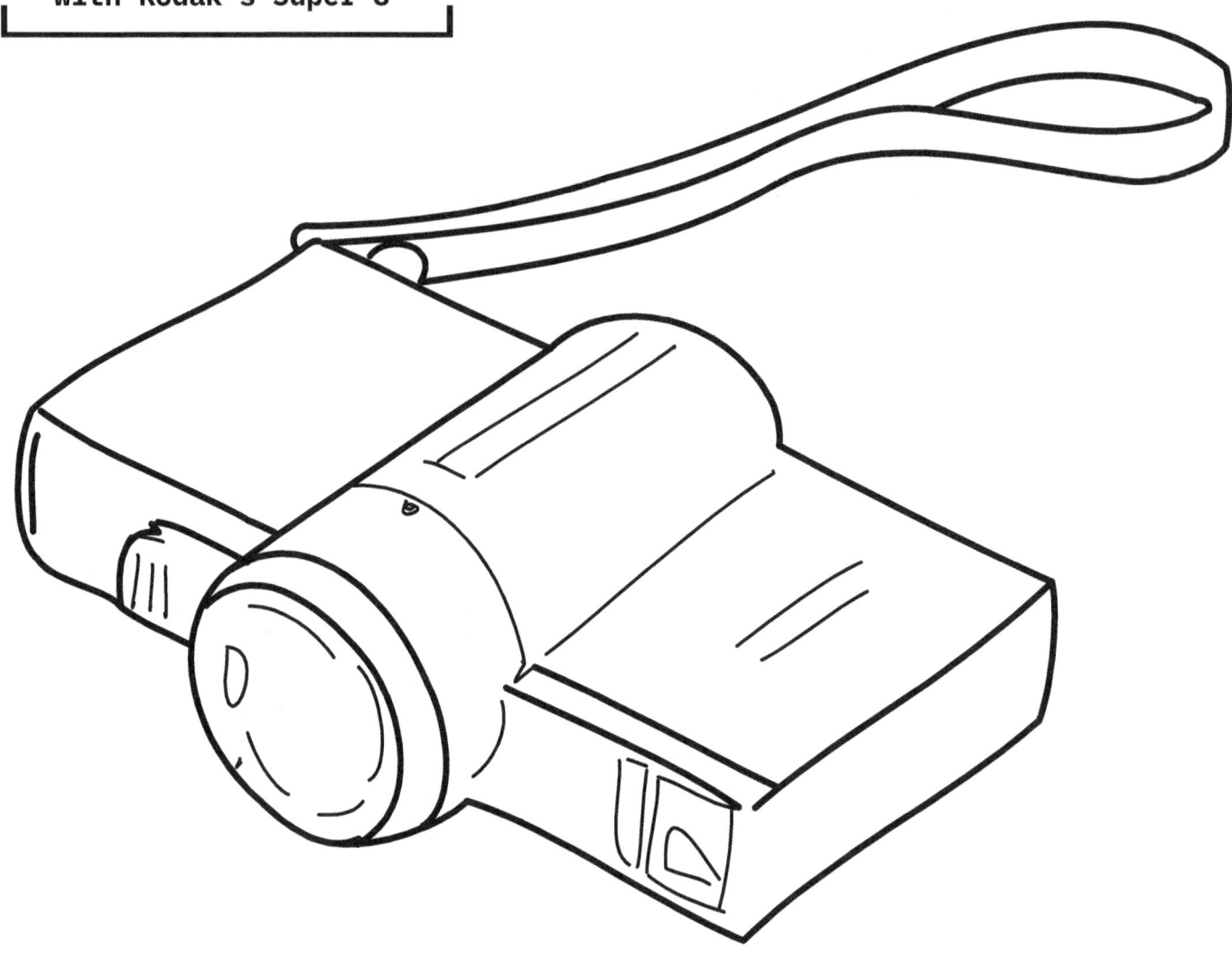

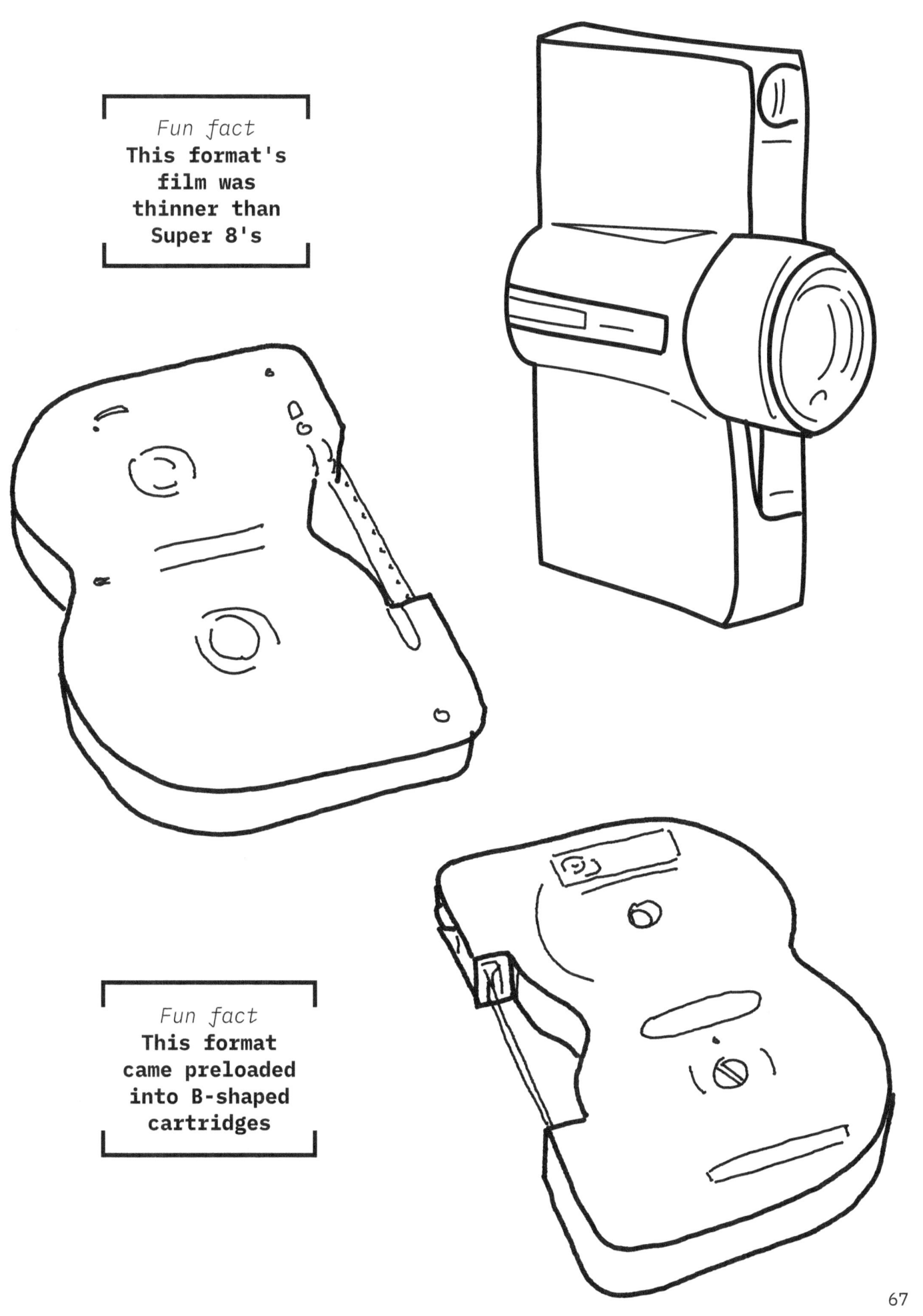

Fun fact
**This format's
film was
thinner than
Super 8's**

Fun fact
**This format
came preloaded
into B-shaped
cartridges**

67

Super 16mm

AKA
Type W

Size
16mm

Aspect Ratio
1.66:1

Developed by
Rune Ericson

Format
analog

Era
1969–

Fun fact
This version of 16mm uses one-sided perforation to have space for a wider image

Fun fact
There is no Super 16mm projector, so playing back this format means making special customizations or enlarging to 35mm

Fun fact
This was an adaptation of the 1.66 aspect ratio ("Paramount format") for 16mm film

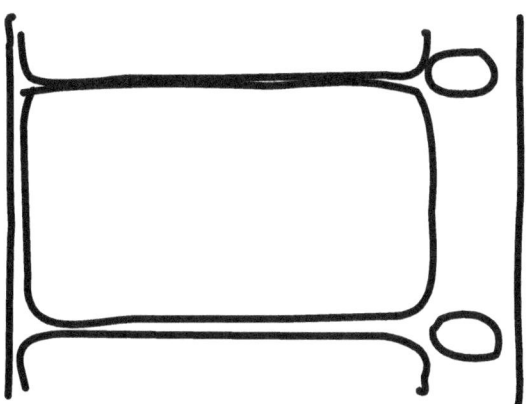

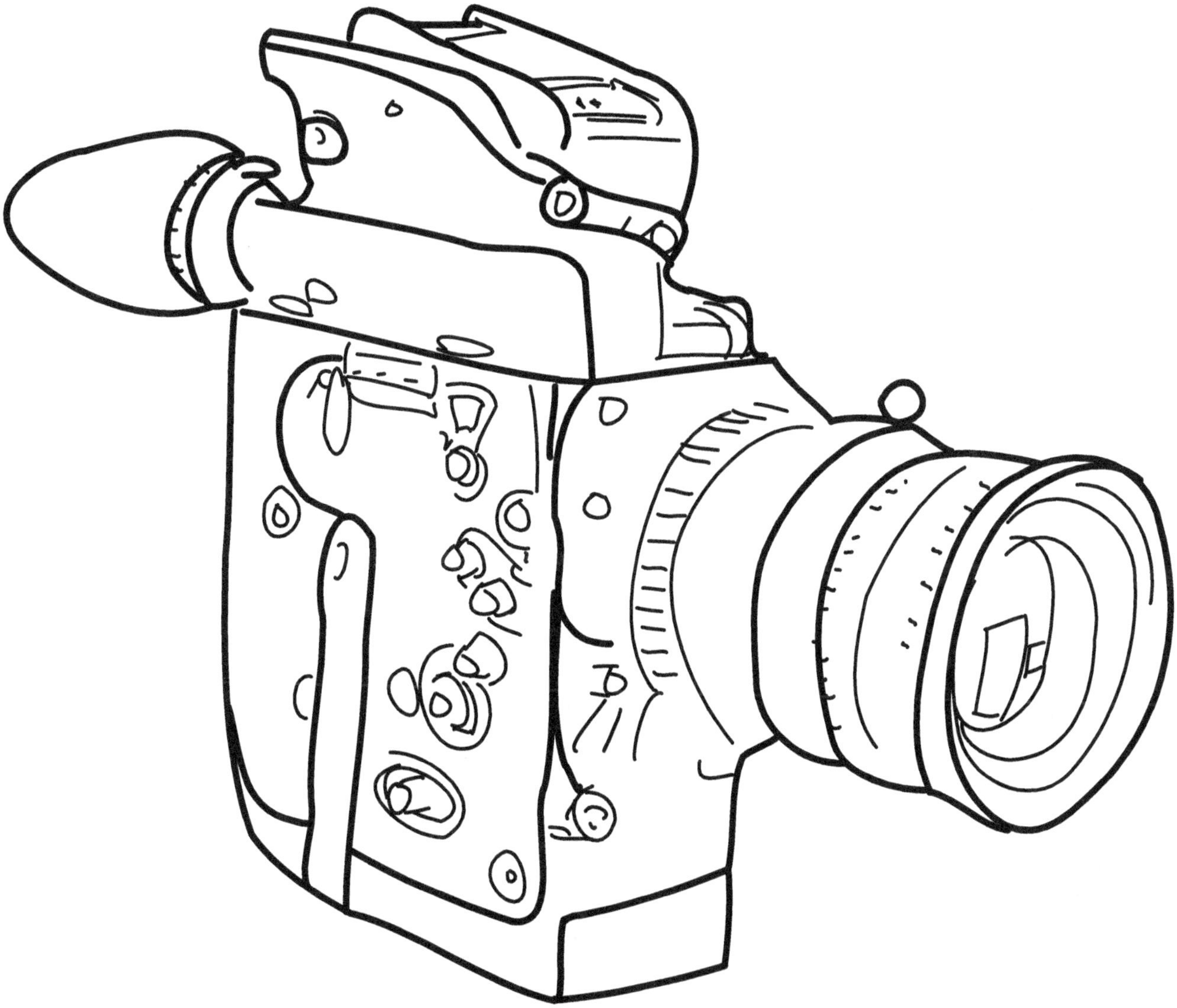

IMAX

Format
analog

AKA
IMAX 15/70

Era
1970-

Aspect Ratio
1.43:1, 1.90:1

Size
70mm (horizontal)

Developed by
IMAX Corporation

Fun fact
**This is a proprietary
system of high-resolution
cameras, film formats,
film projectors,
and theaters**

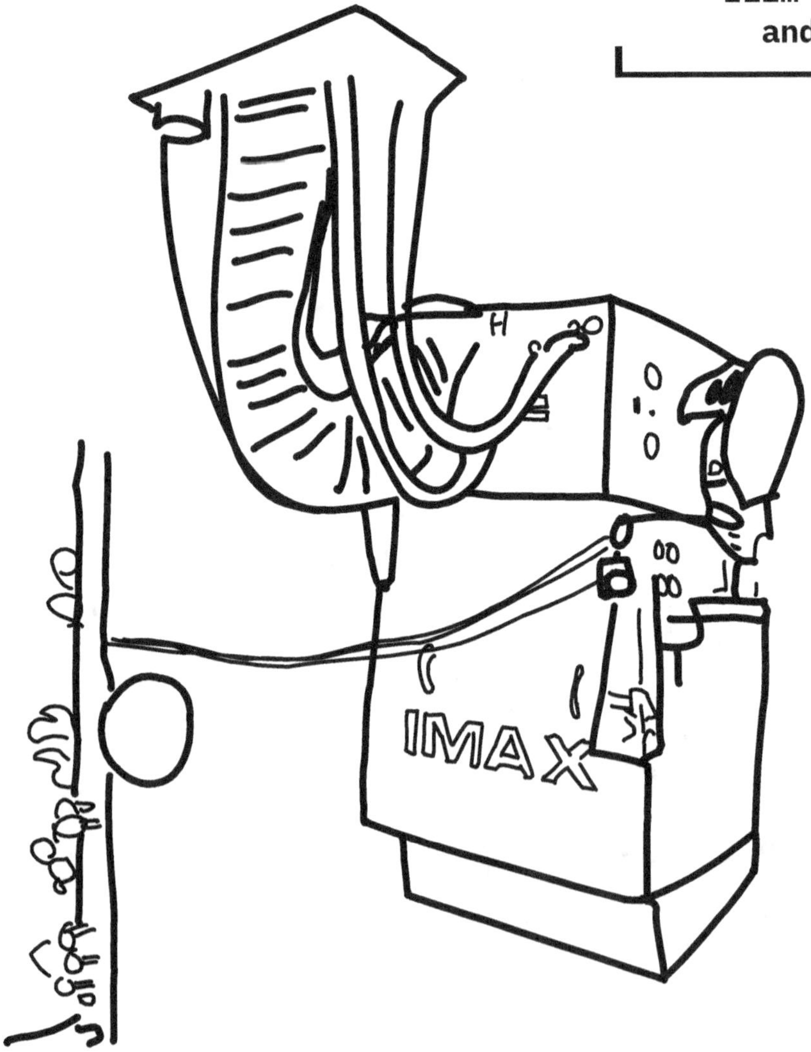

Fun fact
**This horizontal
film version
is called
the 15/70
format, due
to being
70mm and
having 15
perforations
per frame**

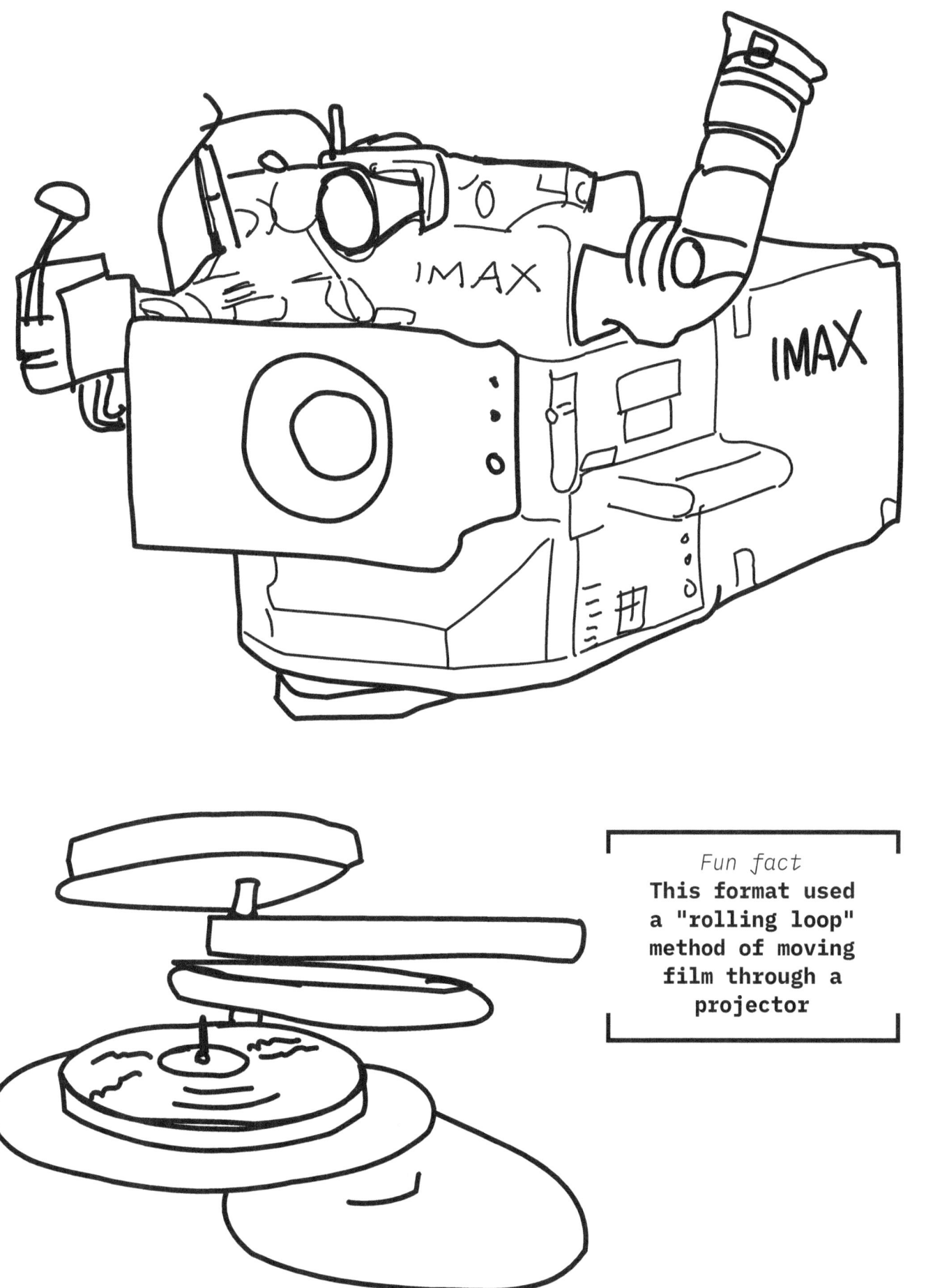

IMAX

IMAX

Omnimax

Format
analog

Era
1976–

Size
70mm (horizontal)

AKA
IMAX Dome

Aspect Ratio
1.43:1

Developed by
IMAX Corporation

Fun fact
This format records with a rounded "fisheye" lens to fit the rounded projection dome

Fun fact
This format requires a custom rounded dome theater, mostly used in planetariums

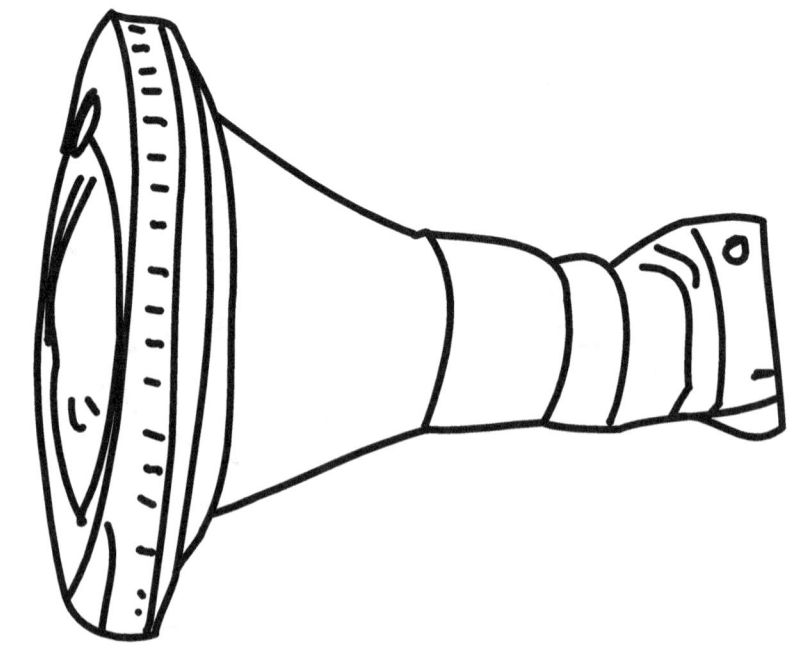

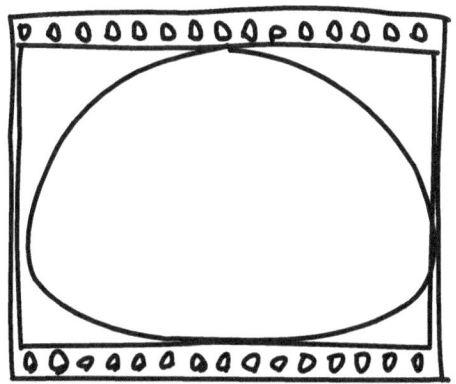

Fun fact
This format's lens was
optically centered
above the film's
horizontal center line

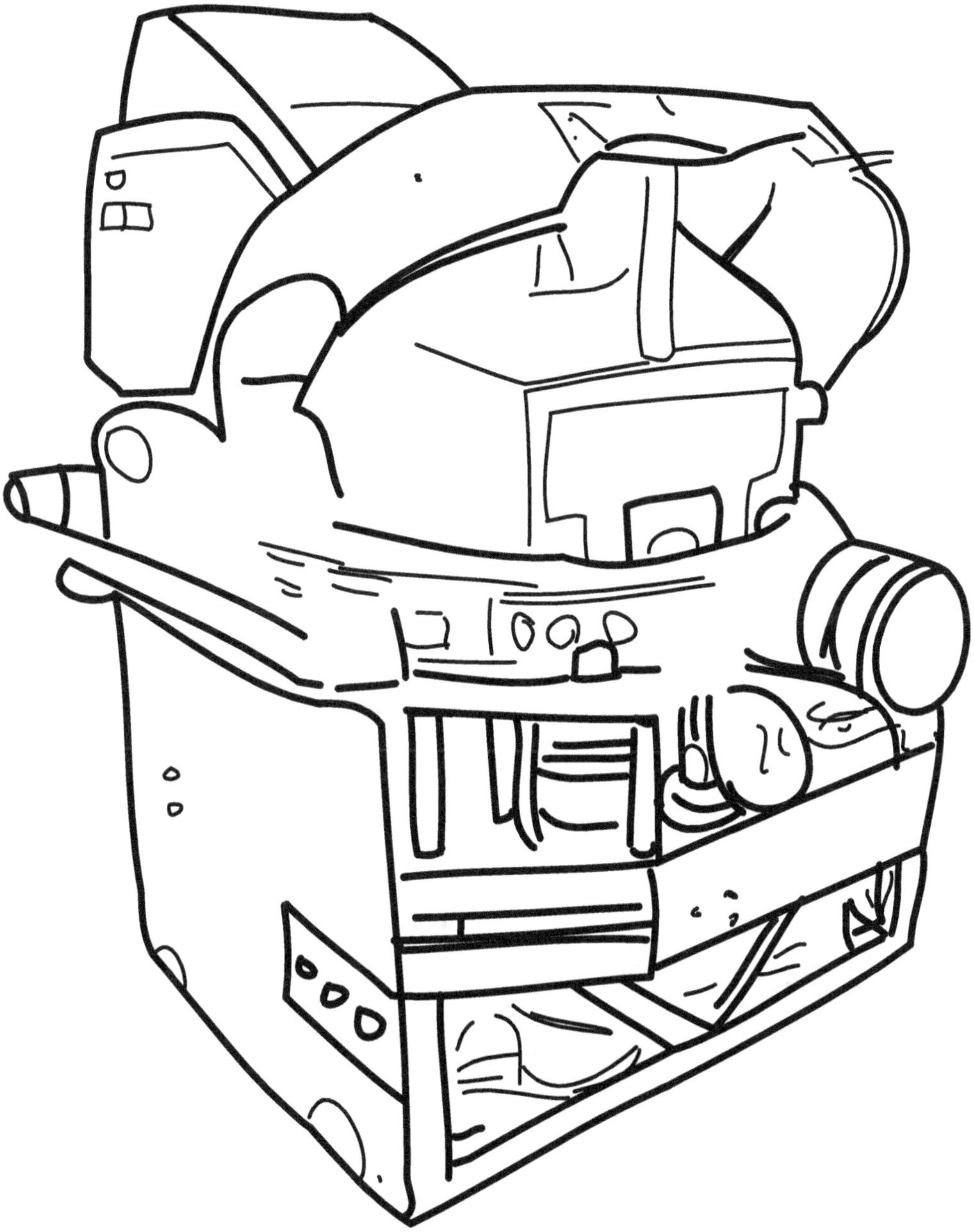

Polavision

Size
8mm

Aspect Ratio
1.36:1

Developed by
Polaroid Corporation

Era
1977-1979

Fun fact
**This format was the Polaroid
Corporation's attempt at bringing
their popular "instant" film to
the moving image market**

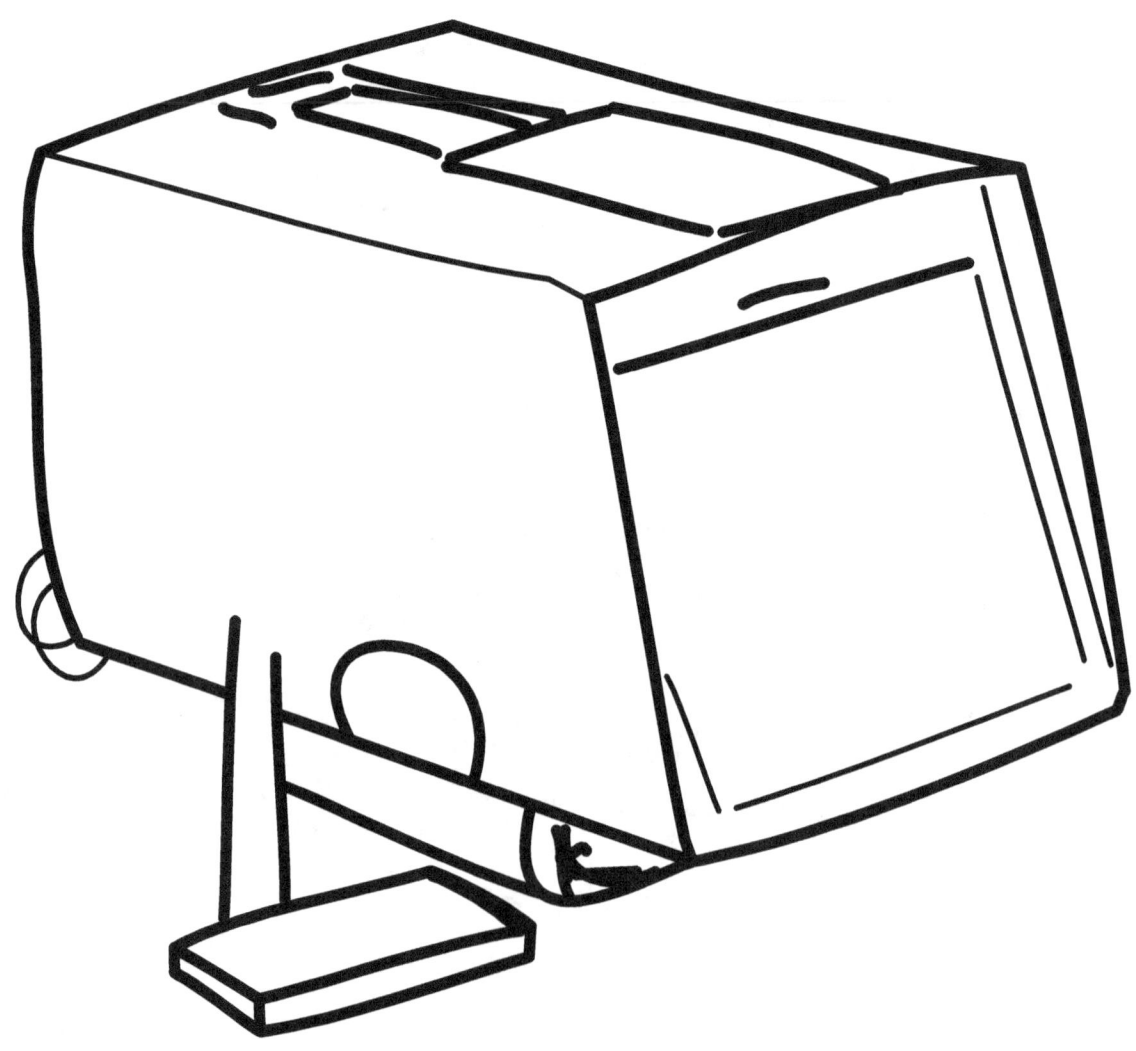

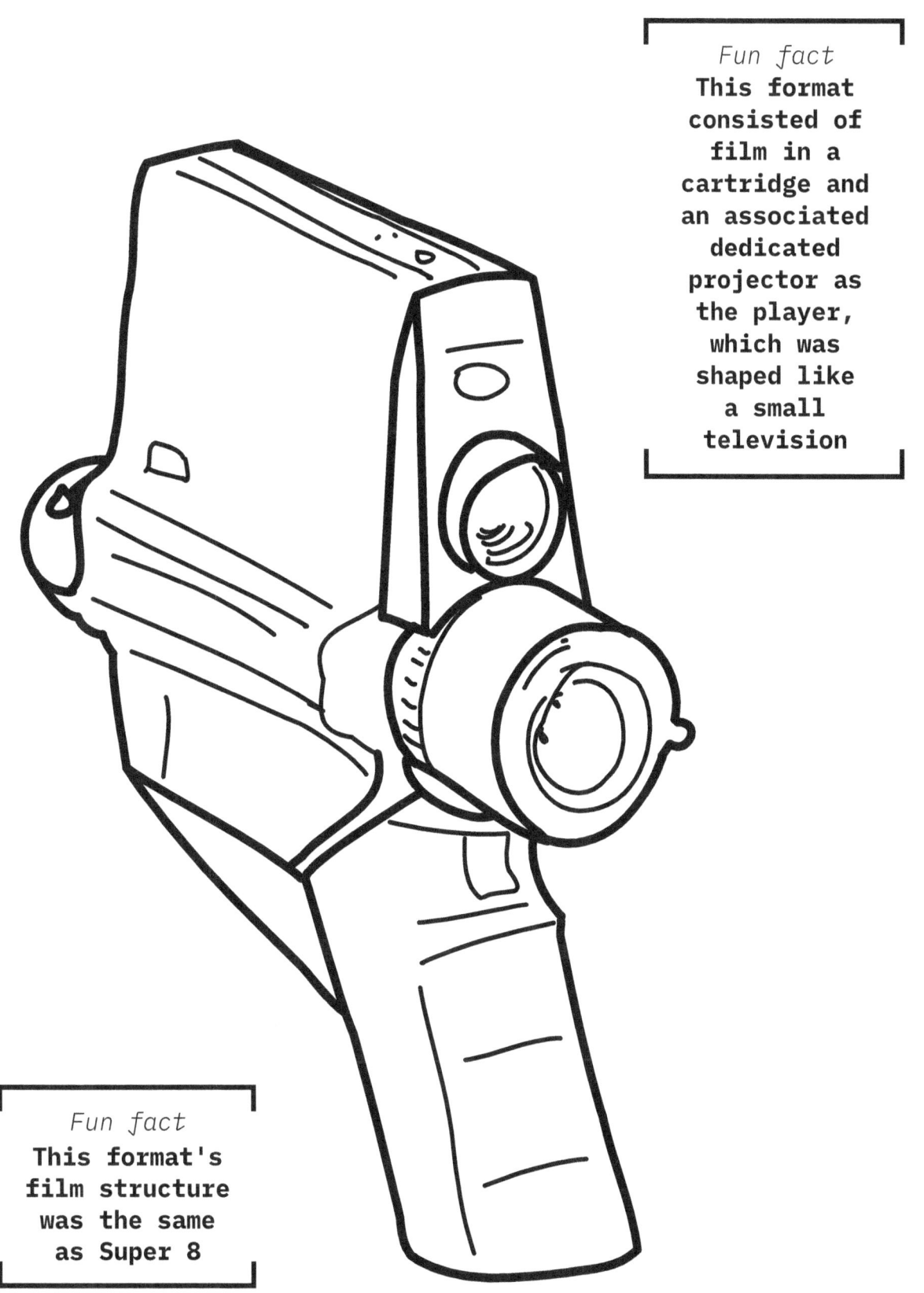

Fun fact
**This format
consisted of
film in a
cartridge and
an associated
dedicated
projector as
the player,
which was
shaped like
a small
television**

Fun fact
**This format's
film structure
was the same
as Super 8**

Showscan

Era
1978–1990s

Developed by
**Douglas Trumbull
(Showscan Film Corporation)**

AKA
CP-65

Size
65mm

Aspect Ratio
2.21

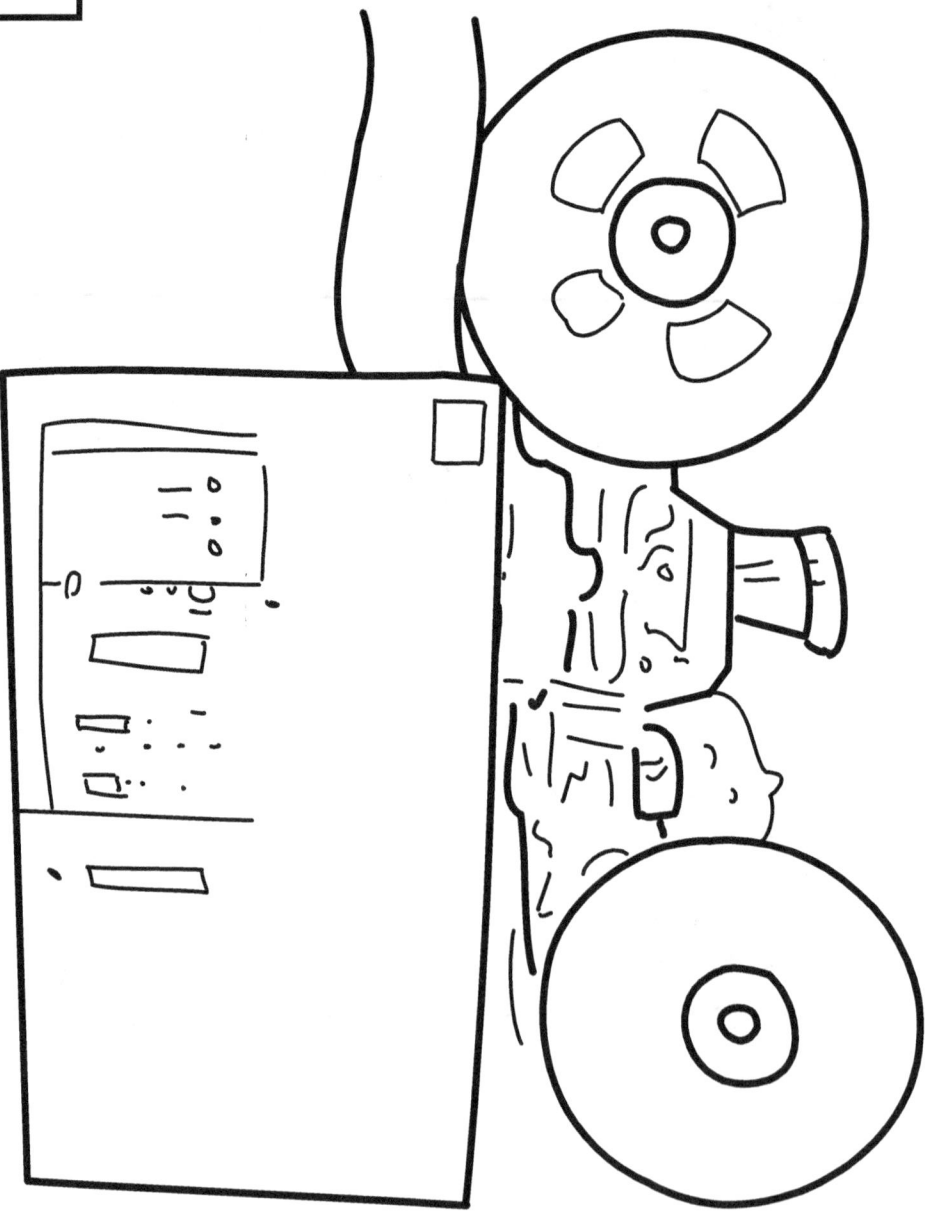

Fun fact
This format failed to compete in the market against IMAX

Fun fact
Douglas Trumbull won an Academy Award for the development of this camera system

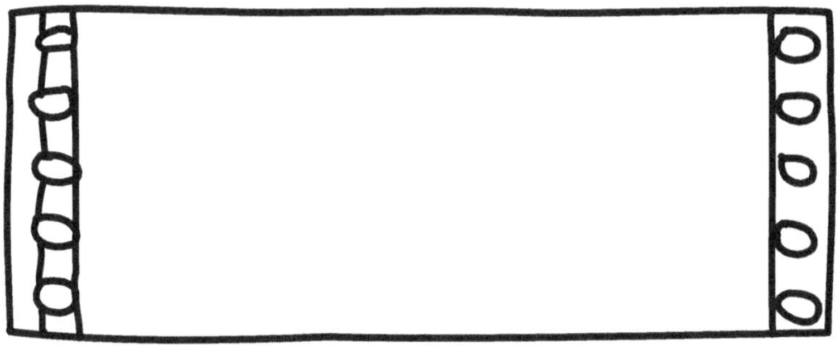

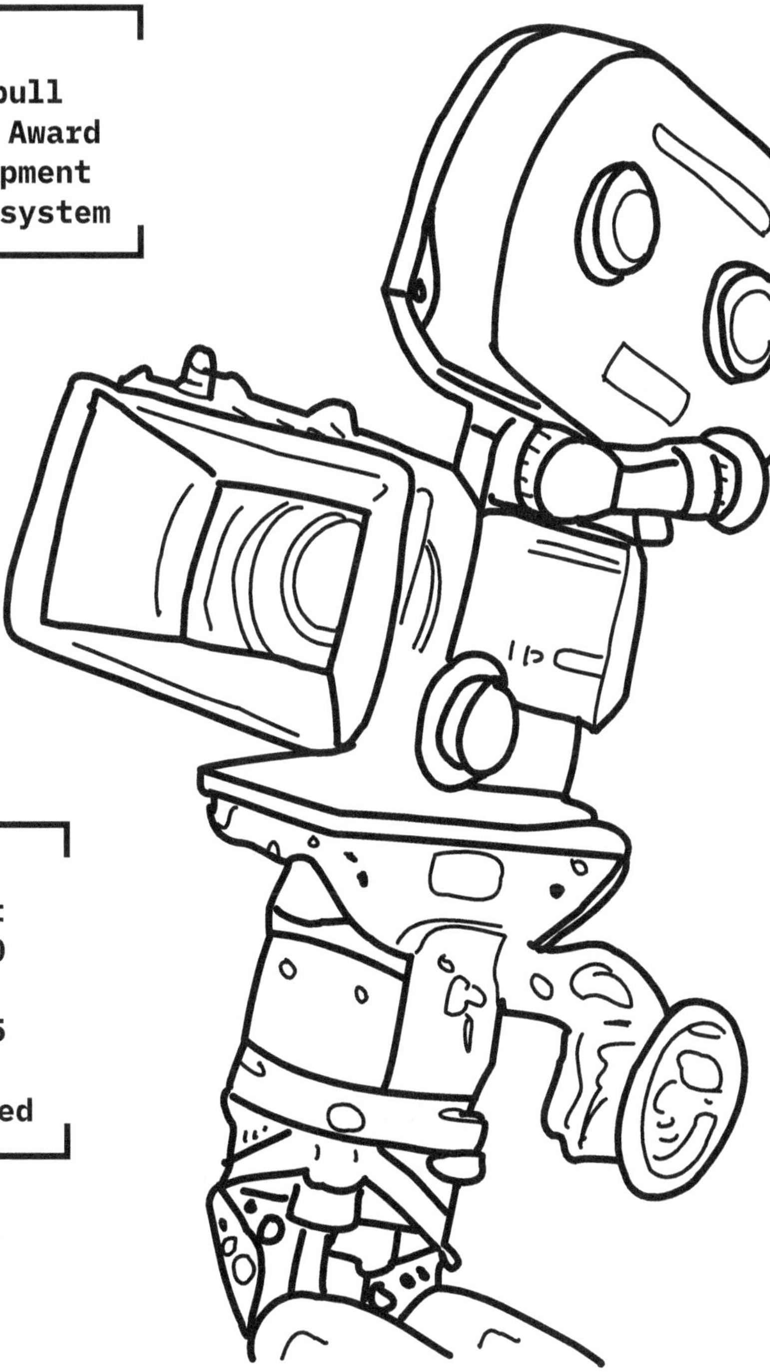

Fun fact
This format moves at 60 frames per second, 2.5 times the standard speed

DCP

Format
digital

AKA
Digital Cinema Package

Size
Various

Aspect Ratio
Various

Fun fact
This format is a structured collection of files that represent a motion picture

Era
2005–

Developed by
Digital Cinema Initiatives, LLC

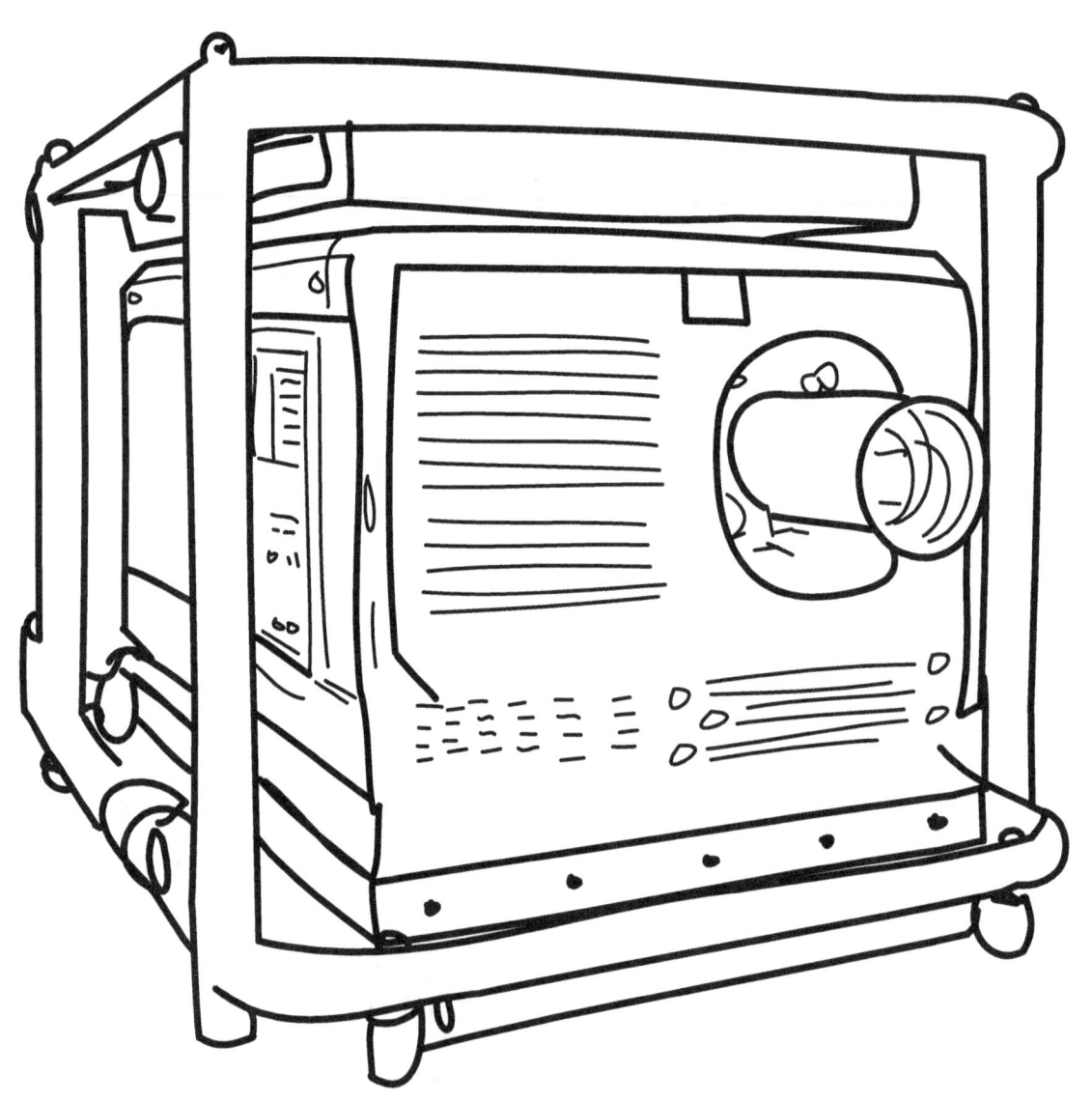

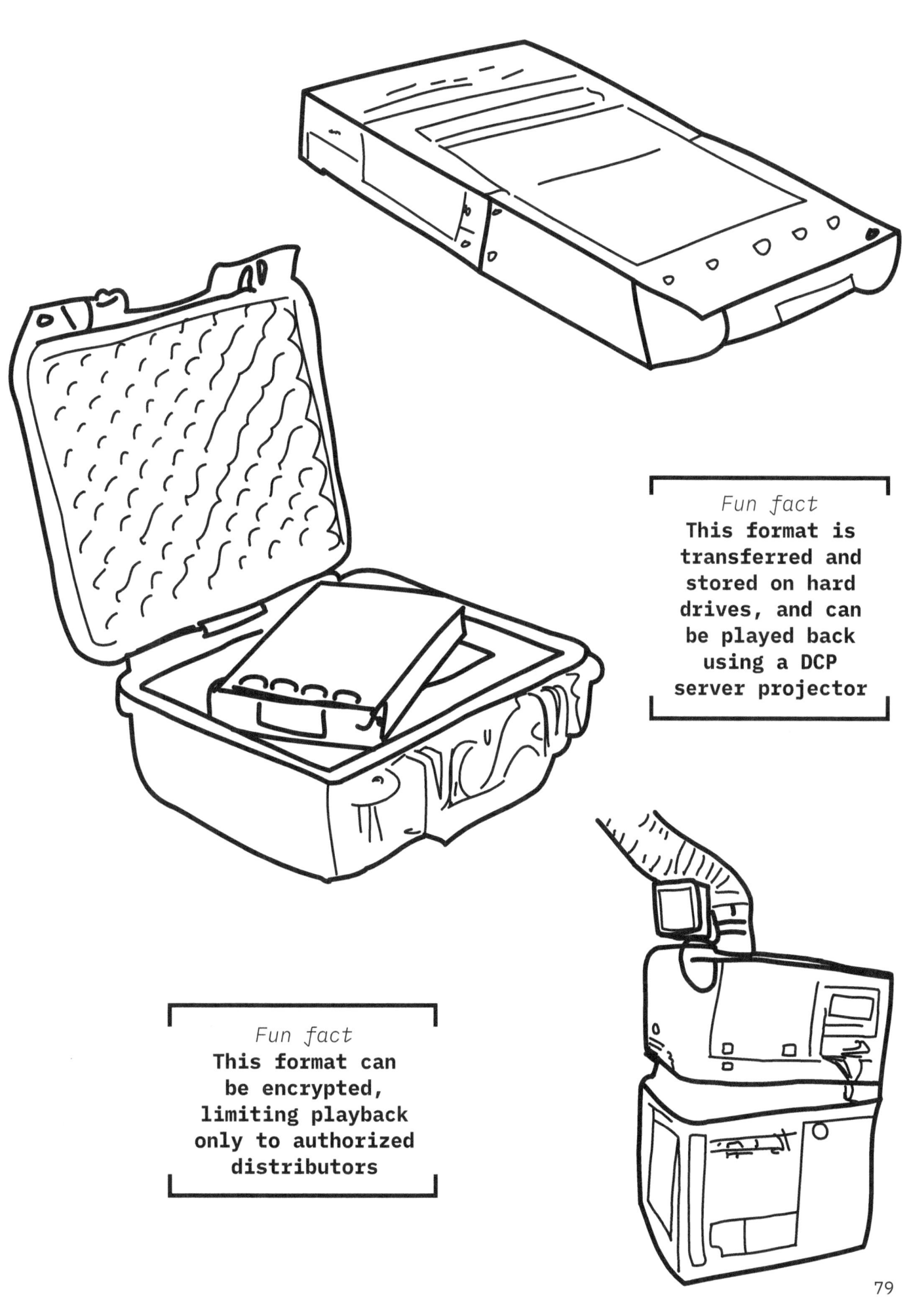

IMAX Laser

Format
digital

Era
2012–

Size
70mm (horizontal)

Developed by
IMAX Corporation

Aspect Ratio
1.43:1

Fun fact
**This format projects
4K-resolution images
using laser technology**

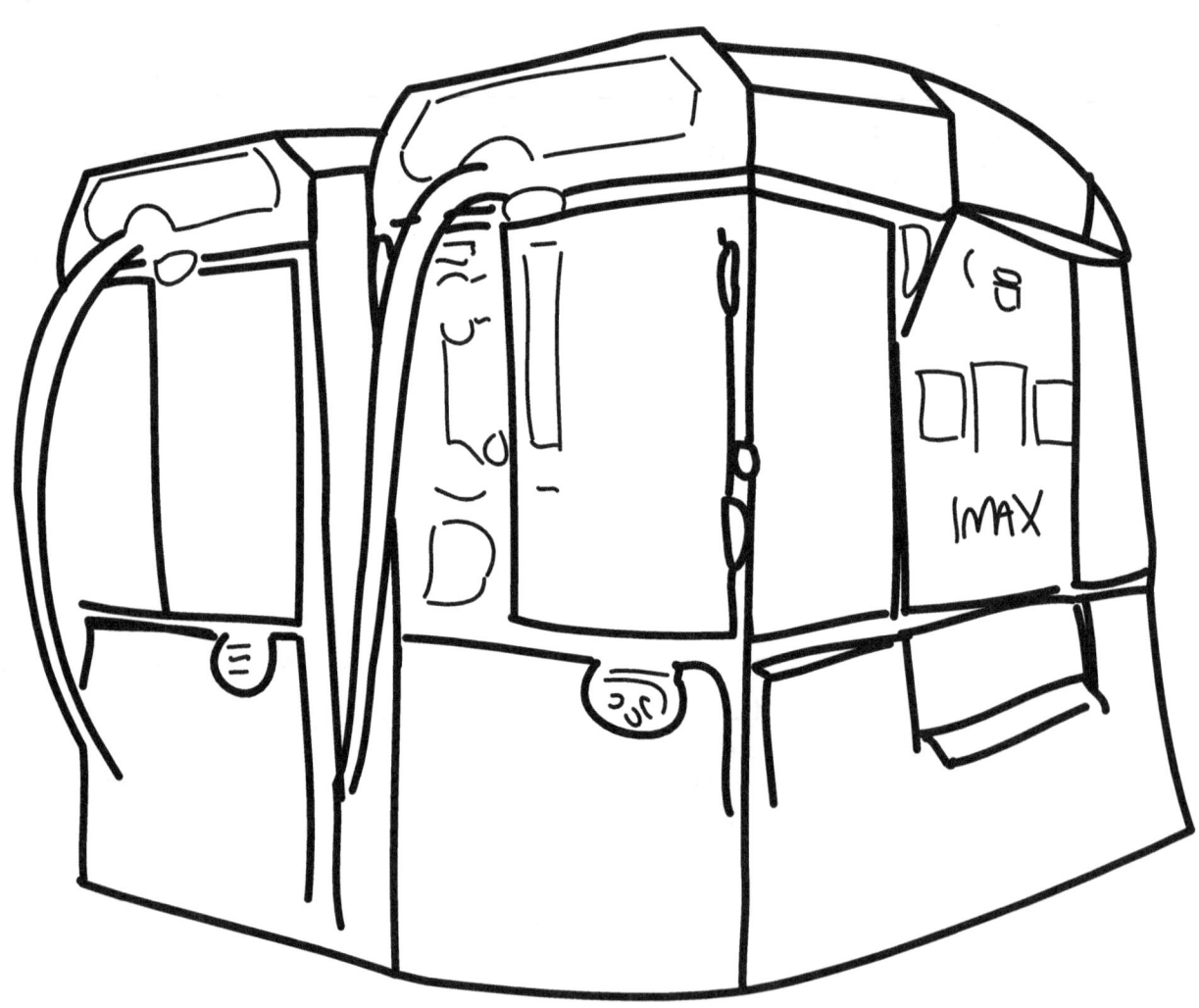

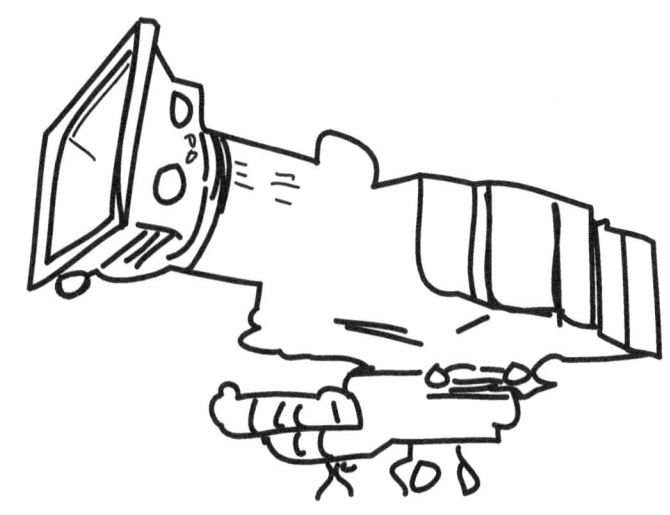

Acknowledgements

Thank you to my technical reviewers, C. Díaz and David Neary, for your deep knowledge and editorial insights.

And yet again thank you to Rory: for endless listening to my hopes and dreams, supporting every one of my ambitions, and for everything.

Also available

The Illustrated Guide to Audio Formats
The Illustrated Guide to Video Formats

About the Author

Ashley Blewer is an archivist,
educator, and software engineer with
over a decade of experience working in
video. Ashley specializes in video and
audio formats, digital preservation,
and communication.

Learn more at
ashleyblewer.com

www.ingramcontent.com/pod-product-compliance
Lightning Source LLC
Chambersburg PA
CBHW081339120626
46546CB00011B/3416